COUNTING

Xing Zhou

Math for Gifted Students

http://www.mathallstar.com

Copyright © 2015 by Xing Zhou. All rights reserved.

No part of this book may be reproduced, distributed or transmitted in any form or by any means, including photocopying, scanning, or other electronic or mechanical methods, without written permission of the author.

To promote education and knowledge sharing, reuse of some contents of this book may be permitted, courtesy of the author, provided that: (1) the use is reasonable; (2) the source is properly quoted; (3) the user bears all responsibility, damage and consequence of such use. The author hereby explicitly disclaims any responsibility and liability; (4) the author is notified in advance; and (5) the author encourages, but does not enforce, the user to adopt similar policies towards any derived work based on such use.

Please visit the website `http://www.mathallstar.com` for more information or email `contact@mathallstar.com` for suggestions, comments, questions and all copyright related issues.

use your mobile device to scan this QR code for more resources including books, practice problems, online courses, and blog.

This book was produced using the LaTeX system.

Contents

Preface   iii

1 Introduction   1

2 Addition and Multiplication Principles   3
    2.1 Counting Principles Explained 3
    2.2 Using the Addition Principle 7
    2.3 Using the Multiplication Principle 10
    2.4 Catches, Catches and Catches 13
    2.5 Factorial 15
    2.6 Practice 18

3 Combination and Permutation   21
    3.1 Does Order Matter? 21
    3.2 The Formulas 24
    3.3 Catches Again 27
    3.4 Practice 29

4 Inclusion-Exclusion Principle (Venn Diagram)   31
    4.1 Set Explained 31
    4.2 Venn Diagram 33
    4.3 Inclusion-Exclusion Principle 36
    4.4 Counting Numbers 39
    4.5 Practice 40

5 Manual Counting   41
    5.1 When to Count Manually 41
    5.2 Systematic Counting 41
    5.3 Tree Analysis 43
    5.4 Hybrid Question 45
    5.5 Practice 48

6 Counting Techniques and Patterns   51
    6.1 Count like a Pro 51

CONTENTS

  6.2 Complementary Counting (Count the Opposite) . . . 54
  6.3 Lattice 56
  6.4 Bundling 58
  6.5 Circle / Round Table 61
  6.6 Permutation of Combination 63
  6.7 Combination of Combination 65
  6.8 Different Balls in Different Boxes 67
  6.9 Derangement 68
  6.10 Mapping 70
  6.11 Symmetry 72
  6.12 Cut the Rope 75
  6.13 Knives and Balls 77
  6.14 Generating Function 78
  6.15 Modeling 80
  6.16 Shortcut 83
  6.17 Practice 85

7 Elementary Probability 89
  7.1 Probability Defined 89
  7.2 Independent and Exclusive Events 90
  7.3 Probability Tree Diagram 95
  7.4 Geometric Probability 99
  7.5 Practice 104

Appendices 109

A Solutions 111
  A.1 *Chapter 1* 112
  A.2 *Chapter 2* 113
  A.3 *Chapter 3* 120
  A.4 *Chapter 4* 123
  A.5 *Chapter 5* 127
  A.6 *Chapter 6* 132
  A.7 *Chapter 7* 139

B Estimate π 149

Preface

Welcome to Math All Star© series!

Math All Star originates from a series of lectures given to a group of gifted middle school students with a love for mathematics and an interest in participating in competitions such as MathCounts, AMC, and AIME. These lectures aim to strengthen their problem-solving abilities and to introduce effective techniques that are not typically taught in the classroom.

As the popularity of Math All Star grew, the author began to upload lecture materials to create online courses, thereby providing students with the opportunity to progress at their own paces.

Since then, course materials have constantly been reviewed and updated to reflect student feedback and the observations made during lectures. Recent competition problems are also continuously analyzed and referenced to ensure the relevance of the contents. These course materials are the foundations of this Math All Star series.

Because competition math is a diversified subject that covers both a wide breadth and depth of topics, it is quite challenging to effectively cover all the material in one book that is appropriate for every interested student. Consequently, the author has decided to write a series of books, with each one focusing on a particular topic. Students are encouraged to pick and choose where to begin, depending on their individual skill levels and needs.

Preface

In addition to these books, the Math All Star website provides extra practice problems and serves as a highly recommended supplemental learning resource.

If there are any questions, comments, or concerns, please visit the website or email contact@mathallstar.com.

Happy learning!

To visit the Math All Star website, scan this QR code or go directly to http://www.mathallstar.com

Chapter 1

Introduction

You think everybody can count? Take a look at some counting problems available on http://www.mathallstar.com and you might just think differently!

Counting problems appear in all levels of math competitions. It is also one type of problems that senior students and even adults may not necessarily do better than well-trained junior students. This is because counting problems usually do not involve complex theorems or formulas. Rather, they demand a systematic and analytical approach. In these cases, a trained brain becomes more important than a head full of formulas.

This book will not only cover the formulas, but also focus on explanations of different techniques that can be used, including the following three principles that are the basis for solving the vast majority of counting problems:

- the Addition Principle
- the Multiplication Principle
- the Inclusion-Exclusion Principle (Venn Diagram)

Chapter 1: Introduction

You must thoroughly understand them in order to become a counting pro.

Additionally, although the permutation and combination formulas are two very important tactical building blocks that are frequently used to tackle counting problems, proper analysis of the questions should always be the first step taken in order to avoid falling for the tricks.

However, knowing just some principles and formulas is far from enough to excel in counting. It is just as important to be familiar with frequently-used techniques and well-known patterns, which are the footholds for future success. That is the reason about half of this book is dedicated to the aforementioned topics. Fully understanding them will significantly improve your problem-solving skills.

As a final note, while both elementary probability and geometric probability are also explored in this book, the discussions are limited to the middle school and high school levels.

Chapter 2

Addition and Multiplication Principles

We will start our journey by learning these two basic counting principles. One cannot claim that one knows counting without mastering these two principles.

2.1 Counting Principles Explained

To understand these two counting principles, let's first look at a simple example.

Chapter 2: Addition and Multiplication Principles

Example 2.1.1

How many different routes can we take from A to B without going backward?

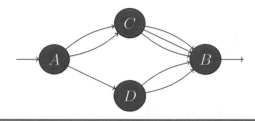

This is an easy question, isn't it? The answer is $2 \times 3 + 1 \times 2 = 8$.

Why does this expression, $2 \times 3 + 1 \times 2$, give us the correct answer? What do the "+" and "×" really mean here from the counting perspective?

The "+" here means that there are two mutually exclusive cases ($A \to C \to B$ and $A \to D \to B$). As a result, the number of total possible routes is the sum of the number of choices in these two cases.

The "×" in each case means that if there are two independent back-to-back steps (e.g. $A \to C$ and $C \to B$ in the first case), the number of total possible routes is the product of the choices in all these steps.

Do you agree?

Chapter 2: Addition and Multiplication Principles

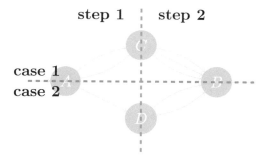

If we formalize the two observations above, we can conclude:

Addition Principle

If one task can be accomplished by completing either one of two **mutually exclusive** cases, the total number of ways to accomplish this task is the sum of the number of ways in these two cases.

and

Multiplication Principle

If a task can be accomplished by completing two **independent** steps in sequence, the total number of ways to accomplish this task is the product of the number of ways in these two steps.

We have utilized both principles in *Example 2.1.1*:

Chapter 2: Addition and Multiplication Principles

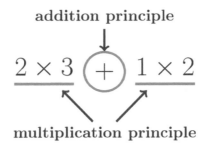

The keywords in the addition principle are *mutually exclusive*. The keyword in the multiplication principle is *independent*.

If either *mutually exclusive* or *independent* is violated, we cannot directly use these principles. For example, if the given route map is modified as in the following, we will need to find another way to solve this new problem.

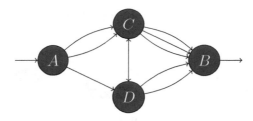

💡 *Tip: The key to solve many counting problems is to list all the mutually exclusive cases and then construct independent steps for each case.*

Exercise 2.1.1

What are the keyword(s) in the addition principle and the multiplication principle, respectively?

2.2 Using the Addition Principle

Let's examine another example:

Example 2.2.1

Randomly draw a card from 1 to 10 twice, with replacement (i.e. the card is put back after each draw). What is the probability that the product of these two cards is a multiple of 7?

We know

$$\text{probability} = \frac{\text{number of the qualified events}}{\text{number of the total events}}$$

In this example, it is easy to count the total possibilities: $10 \times 10 = 100$. (Do you recognize that we are applying the multiplication principle here?)

Therefore the key to solve this problem is to count the qualified events, i.e. when the product of two cards is a multiple of 7.

Solution 1

To have the product equal to a multiple of 7, at least one draw must produce 7. Therefore we list four cases below:

Draw 7?

1st draw	2nd draw	
Yes	Yes	✓
Yes	No	✓
No	Yes	✓
No	No	✗

Chapter 2: Addition and Multiplication Principles

Clearly, these four cases are mutually exclusive and together cover all the possibilities.

Now we can count the total qualified events as:

$$1 \times 1 + 1 \times 9 + 9 \times 1 = 19$$

Therefore the probability is $\frac{19}{100}$.

Done.

Tip: More often than not, a counting problem can have more than one solution. Different solutions usually result from different ways to list cases and to construct steps for each case.

Are there any other solutions to this problem?

The answer is yes.

Solution 2

We note that if the 1st card is 7, it does not matter what the 2nd card is. However, if the 1st card is not 7, the 2nd card must be 7.

Based on this thinking, we list only 3 different cases in the following:

Draw 7?

1st draw	2nd draw	
Yes	*	✓
No	Yes	✓
No	No	✗

8

We can then count qualified events as

$$1 \times 10 + 9 \times 1 = 19$$

The probability is $\frac{19}{100}$, which agrees with our previous result.

Done.

Tip: Counting is one type of problems that you may be able to solve but you may not have confidence about your result. Being able to find multiple solutions and cross check those results can be very useful in boosting your confidence.

It turns out there is another way to solve this problem. Let's observe the cases again:

Draw 7?

1st draw	2nd draw	
Yes	Yes	✓
Yes	No	✓
No	Yes	✓
No	No	✗

Solution 3 (Count the Opposite)

We know these 4 cases cover all possibilities. Three of them lead to qualified events and only one does not. Therefore instead of counting the qualified events, we can count the disqualified events:

$$9 \times 9 = 81$$

As there are a total of 100 different possibilities, the number of qualified events is $100 - 81 = 19$, resulting in the same probability of $\frac{19}{100}$.

Chapter 2: Addition and Multiplication Principles

Done.

As this example has revealed, there can be multiple approaches in solving the same problem. Different approaches may involve different levels of complexity. It is important to employ the most effective approach. *Chapter 6* is devoted to discussion of various counting techniques.

2.3 Using the Multiplication Principle

The three different solutions to the previous example all focus on constructing and choosing *mutually exclusive* cases. This is key in applying the *addition principle.* Now let's look at the *multiplication principle* in detail.

Do you still remember the keyword of the multiplication principle?

independent

What does it mean? It basically means that choices of the previous steps do not affect those of the subsequent steps.

Then the question becomes: what if the steps are not independent? The short answer is we cannot *directly* apply the multiplication principle. The long answer is we have to transform these steps into independent ones, or we cannot use the multiplication principle at all.

There are many techniques that can be used to construct independent steps. The most basic and quite effective way is to determine the sequence of steps.

Let's review some examples:

Example 2.3.1

How many 4-digit numbers can be formed by selecting 4 *different* digits from 0 to 9?

This is a relatively easy question. To emphasize how we should use the multiplication principle, let's go through these steps one by one in detail.

Solution

To make a 4-digit number, we need to complete the following 4 steps:

Step	choose a digit for	number of choices
1	the "thousands" place	9 (excluding "0")
2	the "hundreds" place	9
3	the "tens" place	8
4	the "ones" place	7

Therefore the answer is $9 \times 9 \times 8 \times 7 = 4536$.

Done.

💡 *Tip: when counting, it is important to show your number model, e.g. $9 \times 9 \times 8 \times 7$ in this example. This demonstrates how you work out the solution. Sometimes, if the answer is a big number, it is OK not to do the final calculation.*

Example 2.3.2

Among these 4-digit numbers in the previous example, how many are odd numbers?

🅿 *Pause: think about it before continuing ...*

You may find that the 4 steps above are no longer adequate in solving this example. Why? This is because the example is asking for odd numbers. If we continue using the previous sequence, the choices of the "ones" place will *depend* on how many odd digits have been chosen in steps 1 to 3.

For example, if the first 3 steps have selected an odd digit each time, the "ones" place has only 2 choices left. Otherwise it will have more choices.

Depend is an unwelcome word in counting. Can we somehow transform these steps into independent ones? It turns out all we need in this case is simply to change the sequence of the four steps.

Step	choose a digit for	number of choices
1	the "ones" place	5 (i.e. 1, 3, 5, 7, 9)
2	the "thousands" place	8 (also excluding 0)
3	the "hundreds" place	8
4	the "tens" place	7

Therefore the answer is $5 \times 8 \times 8 \times 7 = 2240$.

Please note that the count of odd numbers is not exactly half of that of the total 4-digit numbers.

❓ *Quiz: can we use a sequence of ones $\to$ tens $\to$ hundreds $\to$ thousands to solve this question?*

2.4 Catches, Catches and Catches

Do you still remember the two keywords when applying the addition and multiplication principles?

They are *mutually exclusive* and *independent*.

In addition to these two principles, there is an important aspect you should watch out for in solving counting problems. It is called the *catches*.

A catch is a constraint that tricks you or prevents you from applying counting principles directly. To solve a challenging counting problem is often the same as to recognize and eliminate catches.

When working on a counting problem, you should always ask yourself:

We will illustrate this point in the following example:

Example 2.4.1

Select 4 different digits from 0 to 9.

(a) How many 4-digit numbers can be formed?

(b) Among the 4-digit numbers, how many are odd numbers?

(c) Among the 4-digit numbers, how many are even numbers?

We have discussed (a) and (b) already. Let's analyze what the catches are in this counting problem:

(a) All 4-digit numbers:

- no leading "0"

(b) Odd numbers:

- no leading "0"

- "ones" place must have an odd digit

(c) Even numbers:

- no leading "0"

- "ones" place must have an even digit

We have already solved (a) and (b). Can we solve (c) by using a similar approach as we do for (a) or (b)?

🔘 *Pause: think about it before continuing ...*

The answer is No. Why? It is because the digit "0" is an even digit itself. As a result, it creates an interdependency between these two catches.

To ensure it is an even number, we need to first choose an even digit for the "ones" place. However, depending on whether "0" is chosen for the "ones" place, the choices for the "thousands" place will vary. In the case of creating an odd number, we do not have this dependency.

Eliminating catches is at the heart of solving counting problems. Almost all counting techniques are developed to help eliminate catches. Let's use the "even number" question as an example.

Solution 1 - Do Casework

We will group them into two cases: one where "0" is selected for the "ones" place, and the other where "0" does not appear in the "ones" place. In both cases, digits are selected in the sequence of

$$"ones" place \to "thousands" place \to "hundreds" \to "tens"$$

case	choices
"0" is chosen for the "ones" place	$1 \times 9 \times 8 \times 7$
"0" is *not* chosen for the "ones" place	$4 \times 8 \times 8 \times 7$

These two cases are mutually exclusive, and cover all possibilities. Therefore the total count is

$$1 \times 9 \times 8 \times 7 + 4 \times 8 \times 8 \times 7 = 2296$$

Done.

This same problem can be solved with an alternative approach:

Solution 2 - Counting the Opposite

We have a total of $9 \times 9 \times 8 \times 7 = 4536$ 4-digit numbers (from Example 2.3.1 on page 11) and $5 \times 8 \times 8 \times 7 = 2240$ odd numbers (from Example 2.3.2 on page 12). Therefore the count of even numbers must be

$$4536 - 2240 = 2296$$

Done.

We will see more examples of counting the opposite in *Chapter 6 Counting Techniques and Patterns*.

2.5 Factorial

You may have noticed that when using the *multiplication principle*, we often need to multiply several consecutive integers. For example: $1 \times 2 \times 3 \times \cdots \times n$. There is a math operator specifically designed for such operation. It is called *factorial*.

Chapter 2: Addition and Multiplication Principles

> **Definition 2.5.1 Factorial**
>
> The factorial of a non-negative integer, n, is written and defined as
> $$n! = 1 \times 2 \times 3 \times \cdots \times n$$

For example:
$$1! = 1$$
$$2! = 1 \times 2 = 2$$
$$3! = 1 \times 2 \times 3 = 6$$
$$4! = 1 \times 2 \times 3 \times 4 = 24$$

A special case, $0! = 1$, is defined by convention.

Exercise 2.5.1

Calculate $5!$.

We can also define $n!$ recursively as following:
$$\begin{cases} 0! = 1 \\ k! = k \times (k-1)! \end{cases}$$

What does this mean? The first line $0! = 1$ defines an initial value. Afterwards we can compute any factorial by using the previous value:

$$0! = 1 \qquad \textit{initial value}$$
$$1! = 1 \times (1-1)! = 1 \times 0! = 1$$
$$2! = 2 \times (2-1)! = 2 \times 1! = 2$$
$$3! = 3 \times (3-1)! = 3 \times 2! = 6$$
$$4! = 4 \times (4-1)! = 4 \times 3! = 24$$
$$\cdots$$

Using a recursive pattern to define function values can be an effective approach. It is the base of a powerful proof method called induction. Furthermore, recursive functions are also used in computer programming.

In addition to the regular factorial that we have just discussed, there is a *double factorial* operator. It is used in situations such as multiplying consecutive odd (or even) numbers. Its definition is as follows:

Definition 2.5.2 Double Factorial

The double factorial of integer, $n \geq -1$, is defined as:
$$n!! = \begin{cases} n \cdot (n-2) \cdot (n-4) \cdots 1 & , \text{ where } n > 0 \text{ and odd} \\ n \cdot (n-2) \cdot (n-4) \cdots 2 & , \text{ where } n > 0 \text{ and even} \\ 1 & , n = -1 \text{ or } 0 \end{cases}$$

For example:
$$3!! = 3 \times 1 = 3$$
$$4!! = 4 \times 2 = 8$$

Chapter 2: Addition and Multiplication Principles

2.6 Practice

Practice 1

Restaurant MAS offers a set menu with 3 choices of appetizers, 5 choices of main dishes, and 2 choices of desserts. How many possible combinations can a customer have for one appetizer, one main dish, and one dessert?

Practice 2

How many factors does 20 have?

Practice 3

Find the number of different rectangles that satisfy the following conditions:

- its area is 2015
- the lengths of all its sides are integers

Practice 4

Eight chairs are arranged in two equal rows. Joe and Mary must sit in the front row. Jack must sit in the back row. How many different seating plans can they have?

Practice 5

Two Britons, three Americans, and six Chinese form a line:

(a) How many different ways can the 11 individuals line up?

(b) If two people of the same nationality cannot stand next to each other, how many different ways can the 11 individuals line up?

Practice 6

Use digits 1, 2, 3, 4, and 5 without repeating to create a number.

(a) How many 5-digit numbers can be formed?

(b) How many numbers will have the two even digits appearing between 1 and 5? (e.g.12345)

Practice 7

Joe plans to put a red stone, a blue stone, and a black stone on a 10×10 grid. The red stone and the blue stone cannot be in the same column. The blue stone and the black stone cannot be in the same row. How many different ways can Joe arrange these three stones?

Practice 8

How many different 6-digit numbers can be formed by using digits 1, 2, and 3, if no adjacent digits can be the same?

Chapter 2: Addition and Multiplication Principles

Practice 9

How many different ways are there to arrange 3 black balls and 3 white balls in a circle? Two arrangements are considered the same if they are different just by rotating the balls.

Practice 10

Joe wants to write 1 to n in a $1 \times n$ grid. The number 1 can be written in any grid, while the number 2 must be written next to 1 (can be at either side) so that these two numbers are together. The number 3 must be written next to this two-number block. This process goes on. Every new number written must stay next to the existing number block.

How many different ways can Joe fill this $1 \times n$ grid?

Chapter 3

Combination and Permutation

Some people think to learn counting is the same as to learn the combination and permutation formulas. While it is certainly an incomplete understanding, it does correctly point out the importance of these formulas as building blocks to solve counting problems.

3.1 Does Order Matter?

Order turns out to be a key factor to consider in counting, as illustrated in the following two examples.

Example 3.1.1

How many 4-digit numbers can be formed by selecting 4 different digits from 1 to 9?

Chapter 3: Combination and Permutation

Example 3.1.2

How many possibilities do we have by selecting 4 different digits from 1 to 9?

First, what is the difference between these two questions?

⏸ *Pause: think about it before continuing ...*

The difference is whether order matters. Let's say we have chosen 1, 2, 3, and 4. With these 4 digits, we can form many different 4-digit numbers by changing their order. However in Example 3.1.2, these 4 digits just represent one possibility because order does not matter.

Therefore in addition to the keyword "catch", you should also remember a new keyword "order". This means you should always ask yourself the following question when working on a counting problem:

As you will notice later, if order matters, it is a permutation question. Otherwise, if order does not matter, it is a combination question. They are two basic building blocks of many counting problems.

For convenience, let's rename:

- Example 3.1.1 as Question P, where order matters

- Example 3.1.2 as Question C, where order does not matter

Chapter 3: Combination and Permutation

Based on the *multiplication principle*, it is straightforward to solve Question P. We have 9 choices for the first digit, 8 choices for the second digit, 7 choices for the third digit, and 6 choices for the fourth digit. Therefore the number of the total choices is $P = 9 \times 8 \times 7 \times 6$.

How about Question C then? The key to solve Question C is to build a mapping with Question P. Every selection in Question C is mapped to a fixed number of selections in Question P.

To see this, let's still use $\{1, 2, 3, 4\}$ as an example. This set can only count as one selection in Question C. The question is how many selections in Question P it is mapped to. To state it differently, we are trying to determine the count of 4-digit numbers that can be formed by using $\{1, 2, 3, 4\}$. By the *multiplication principle*, the answer is $4 \times 3 \times 2 \times 1 = 24$. This means one selection of $\{1, 2, 3, 4\}$ in Question C is mapped to 24 selections in Question P.

This relationship is shown in the diagram below:

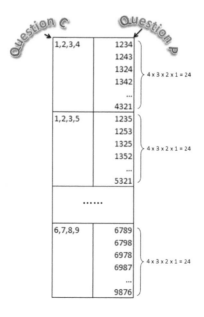

Based on this mapping, we can solve Question C as:

$$C = \frac{P}{4 \times 3 \times 2 \times 1} = \frac{9 \times 8 \times 7 \times 6}{4 \times 3 \times 2 \times 1}$$

The denominator can be seen as the number of repeated objects. So the essence of this mapping technique is to divide the count of all possibilities by the count of duplicates.

> *Dividing by the count of duplicates is a useful technique. We will use this later to solve some more challenging problems. Make sure you fully understand it before continuing.*

3.2 The Formulas

Now we are ready to define permutation and combination.

> **Definition 3.2.1 Permutation**
>
> The act of rearranging k elements from n possible choices ($n \geq k$) into a specific order, without repetition, is called a permutation.

Permutation is usually written as P_n^k, $P(n,k)$, or A_n^k. By the *multiplication principle*:

$$P_n^k = n \cdot (n-1) \cdots (n-k+1) = \frac{n!}{(n-k)!} \qquad (3.1)$$

In particular, $P_n^n = n!$.

Question P is a permutation problem. Selecting 4 digits from 9 choices (from 1 to 9) can form $P_9^4 = 9 \times 8 \times 7 \times 6 = \frac{9!}{(9-4)!}$ different 4-digit integers.

Chapter 3: Combination and Permutation

The count of the 4-digit numbers that can be formed by using 1, 2, 3, and 4 is also a permutation question. The answer is $P_4^4 = 4! = 4 \times 3 \times 2 \times 1$.

> **Definition 3.2.2 Combination**
>
> A combination means to select k elements from n possible choices $(n \geq k)$ in such a way where the order of selection does not matter.

Combination is usually written as C_n^k, $C(n, k)$, or $\binom{n}{k}$. Using the mapping technique discussed earlier leads to the following formula:

$$C_n^k = \frac{P_n^k}{P_k^k} = \frac{n \cdot (n-1) \cdots (n-k+1)}{k \cdot (k-1) \cdots 1} = \frac{n!}{(n-k)! \cdot k!} \qquad (3.2)$$

Intuitively, you can think of permutation as forming a line and combination as forming a set.

Combination has an important and useful property:

> **Property 3.2.1 Combination Property**
>
> $$C_n^k = C_n^{n-k}$$

Let's first evaluate an example:

Example 3.2.1

Calculate C_{100}^{98}.

We can certainly apply its definition:

$$C_{100}^{98} = \frac{100 \times 99 \times 98 \times 97 \times \cdots \times 4 \times 3}{1 \times 2 \times 3 \times \cdots \times 97 \times 98}$$

Chapter 3: Combination and Permutation

This expression can be simplified by canceling common factors between its numerator and denominator. This will lead to:

$$C_{100}^{98} = \frac{100 \times 99 \times \cancel{98} \times \cancel{97} \times \cdots \times \cancel{4} \times \cancel{3}}{1 \times 2 \times \cancel{3} \times \cdots \times \cancel{97} \times \cancel{98}} = \frac{100 \times 99}{1 \times 2}$$

We also have:

$$C_{100}^{2} = \frac{100 \times 99}{1 \times 2}$$

Therefore it seems that C_{100}^{98} indeed is equal to C_{100}^{2}. But why?

▮ *Pause: think about it before continuing ...*

The reason is simple: choosing k elements is equivalent to choosing its opposite $(n-k)$ elements. For example: if there are 10 students in a room, choosing 7 to leave is the same as choosing 3 to stay.

This argument essentially uses the concept of symmetry. Symmetry is a geometric concept, but can be used in many other areas. In the context of counting, being symmetrical means there is no difference between counting A and counting B. In the example above, there is no difference between selecting k elements and selecting $(n-k)$ elements from the same set of n elements.

Symmetry turns out to be a very useful and powerful technique. We will discuss it further in *Section 6.11 Symmetry* later.

❓ *Quiz: does $P_n^k = P_n^{n-k}$ hold?*

Exercise 3.2.1

Calculate C_{1000}^{998}.

3.3 Catches Again

What should one watch out for when working with the permutation and combination formulas? It is still the same keyword: catch!

In math competitions, do not start solving counting problems by directly applying these two formulas. One should always begin analyzing a problem by checking potential catches to avoid traps.

If there are catches, try to eliminate them before applying these two formulas. Let's evaluate a couple of examples.

Example 3.3.1

How many 4-digit numbers can be formed by using 4 different digits from 1 to 9?

No catch. It's a permutation problem. Therefore the answer is P_9^5.

Chapter 3: Combination and Permutation

Example 3.3.2

How many 4-digit numbers can be formed by using 4 different digits from 0 to 9?

Watch out! The catch is no leading "0". Therefore we cannot directly apply the permutation formula or give the answer as P_{10}^4.

Solution 1

Applying the multiplication principle, and calculate the answer as $9 \times 9 \times 8 \times 7 = 4536$.

Solution 2

Let's first transform this problem into one that is free of catches, and then employ the count the opposite technique (also referred to as complementary counting) to get the final answer:

- First, let's allow leading "0". The answer is P_{10}^4

- Next, let's count numbers with leading "0". This is equivalent to forming a 3-digit number by using 1 - 9 only. The answer is P_9^3

Therefore the answer is $P_{10}^4 - P_9^3 = 5040 - 504 = 4536$.

Both solutions provide the same answer.

3.4 Practice

Practice 1

Ten people attend a meeting. Each person shakes hands with the other attendees exactly once. How many times do they shake hands in total?

Practice 2

n teams participate in a friendly tournament, in which each team plays against all the other teams exactly once. If there are a total of 36 games played, find n.

Practice 3

There are 8 different points on a circle. How many triangles can be formed by using these points as vertices?

Practice 4

How many terms are there if we fully expand the following polynomial?
$$(a_1 + b_1)(a_2 + b_2) \cdots (a_n + b_n)$$

Practice 5

Eight couples attend a party. Every gentleman will shake hands with all the other guests except his own wife. Ladies do not shake hands with other ladies. Find out the total number of handshakes.

Chapter 3: Combination and Permutation

Practice 6

n couples attend a ceremony and sit in a row. If every couple must sit together, how many different seating plans are there?

Practice 7

In a soccer game, team MAS has 11 players on the field now and 7 substitutes as backup.

Per rules, at most 3 substitutions can be made in any game. A player that is replaced by a substitute cannot go back to the field during that game.

If at the end of this game, team MAS still has 11 players on the field, how many different possibilities are there?

Chapter 4

Inclusion-Exclusion Principle (Venn Diagram)

The *inclusion-exclusion principle* is used to count when cases are *not* mutually exclusive. It often appears together with a Venn diagram. A Venn diagram is a simple but powerful tool to visualize the *inclusion-exclusion principle*.

4.1 Set Explained

One good way to start learning the inclusion-exclusion principle as well as the Venn diagram is to understand a new concept: *set*.

Set encompasses a complete theoretical system with rich contents. In this chapter, we will focus on counting elements in a *set*.

So what is a *set*?

Chapter 4: Inclusion-Exclusion Principle (Venn Diagram)

> **Definition 4.1.1 Set**
>
> A set is a collection of elements. Elements in a set are *unique* and *un-ordered*.

The keywords here are: *unique* and *un-ordered*.

Following are a few set examples:

- $A = \{1, 2, 3, 4, 5\}$ is a set of five elements
- $B = \{5, 6, 7, 8, 9\}$ is also a set of five elements
- $C = \{5, 4, 3, 2, 1\}$ is the same set as A (Why?)

It is possible for a set to contain unlimited numbers of elements, such as:

- The set of all integers is written as $\mathbb{Z}$
- The set of all real numbers is written as $\mathbb{R}$

If an element k belongs to a set A, we write this relationship as $k \in A$. Otherwise we write it as $k \notin A$. With the concept of a set, we can re-write a statement like "let n be an integer" as "let $n \in \mathbb{Z}$".

There are several set operators. Here, we are most interested in three of them:

- Count (how many elements in a set), written as $|A|$
- Union (merge two sets), written as $A \cup B$
- Intersection (find common elements between two sets), written as $A \cap B$

Let's evaluate an example:

Example 4.1.1

Let $A = \{1,2,3,4,5\}$ and $B = \{5,6,7,8,9\}$, then

- $|A| = 5$, $|B| = 5$
- $A \cup B = \{1,2,3,4,5,6,7,8,9\}$
- $|A \cup B| = 9$
- $A \cap B = \{5\}$
- $|A \cap B| = 1$

From the counting perspective, we are most interested in counting the number of elements in a merged set, which is a union of two or more sets.

When there are only two sets involved, especially when actual elements of both sets are given, it is usually not difficult to work out the count by listing all the elements in the merged set. This is demonstrated in the example above. Our objective, however, is to derive a general rule that works with all situations whether or not actual elements are given.

4.2 Venn Diagram

First, let's review *Example 4.1.1*:

Chapter 4: Inclusion-Exclusion Principle (Venn Diagram)

Let $A = \{1, 2, 3, 4, 5\}$ and $B = \{5, 6, 7, 8, 9\}$, then

- $A \cup B = \{1, 2, 3, 4, 5, 6, 7, 8, 9\}$
- $A \cap B = \{5\}$

From the perspective of element count, we have:

- $|A| = 5$
- $|B| = 5$
- $|A \cup B| = 9$
- $|A \cap B| = 1$

Both A and B have 5 elements, but why does the merged set $A \cup B$ have only 9 elements instead of 10? The reason is: A and B have a common element 5. As elements in a set need to be *unique*, this common element can only be counted once. By definition, all common elements are contained in their intersection $A \cap B$. Therefore we have the following formula:

$$|A \cup B| = (|A|+|B|)-|A \cap B| \tag{4.1}$$

Despite its simplicity, *Equation 4.1* turns out to be an important and frequently used formula.

> Tip: when you see the words "frequently used", you should remember the corresponding formula by heart.

Example 4.2.1

Among all positive integers not exceeding 1000, how many are multiples of either 2 or 5?

Solution

Chapter 4: Inclusion-Exclusion Principle (Venn Diagram)

Let A be the set of all multiples of 2, not exceeding 1000.

Let B be the set of all multiples of 5, not exceeding 1000.

Then $A \cap B$ is the set of all multiples of 10, not exceeding 1000.

Therefore $|A \cup B| = |A| + |B| - |A \cap B| = 500 + 200 - 100 = 600$.

Done.

Equation 4.1 can handle the case of merging 2 sets. We can also derive similar formulas for merging 3 and more sets. However, while developing the formula for the 2-set case is quite intuitive, working out a case involving more than 2 sets will be a bit more involved. (Try it yourself.)

We need a tool to help us visualize and logically derive formulas that involve more than 2 sets. This is where a Venn diagram comes to rescue.

Here is a 2-set Venn diagram:

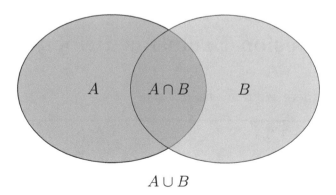

$A \cup B$

In a Venn diagram:

- Each ellipse represents a set

Chapter 4: Inclusion-Exclusion Principle (Venn Diagram)

- The overlapping area represents the intersection of corresponding sets

- The combined area represents the union of these sets

By looking at this diagram, it is clear that the overlapping area (i.e. $A \cap B$) is double-counted when we add A and B. Therefore it must be subtracted when calculating the combined area (i.e. $A \cup B$).

This gives us the same conclusion as *Equation 4.1* on *page 34*:

$$|A \cup B| = (|A|+|B|)-|A \cap B|$$

In other words, when merging two sets A and B, we will need to:

- include: both A and B

- exclude: $A \cap B$

So the essence of the Venn diagram is to visually identify what to include and what to exclude when merging sets, and help us understand the *inclusion-exclusion principle*.

4.3 Inclusion-Exclusion Principle

What if there are more than 2 sets?

Example 4.3.1

Draw a 3-set Venn diagram and derive the corresponding formula.

⏸ *Pause: think about it before continuing ...*

Here is a 3-set Venn diagram:

Chapter 4: Inclusion-Exclusion Principle (Venn Diagram)

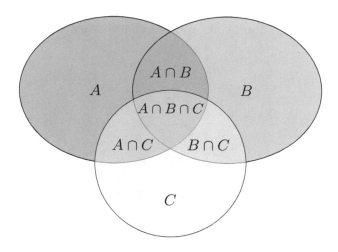

The corresponding formula is:

$$\begin{aligned} & A \cup B \cup C \\ =& (A + B + C) \\ & -(A \cap B + B \cap C + C \cap A) \\ & + A \cap B \cap C \end{aligned} \quad (4.2)$$

Why does Formula 4.2 look like this? We note:

- The three areas, $A \cap B$, $B \cap C$ and $C \cap A$, all cover the center piece, $A \cap B \cap C$

- When we initially perform $A + B + C$, the center piece has been counted 3 times

- When we subtract $A \cap B$, $B \cap C$ and $C \cap A$, we have subtracted the center piece 3 times (This essentially leaves an "empty hole" in the diagram. Can you visualize it?)

- Therefore we must add the center piece back.

Chapter 4: Inclusion-Exclusion Principle (Venn Diagram)

Do you see how inclusion and exclusion work here?

$A \cup B \cup C$
$= (A + B + C)$ *inclusion*
$- (A \cap B + B \cap C + C \cap A)$ *exclusion*
$+ A \cap B \cap C$ *inclusion*

This inclusion and exclusion rule above can be extended to merging n sets:

- There are n blocks
- The 1th block is always the sum of all sets
- The 2nd block is always the sum of the intersections of <u>all</u> possible 2 sets
- The 3rd block is always the sum of the intersections of <u>all</u> possible 3 sets
- ...
- The $(n-1)^{th}$ block is always the sum of the intersections of <u>all</u> possible $(n-1)$ sets
- The nth block is always the intersections of the n sets
- The signs before these blocks **alternate** between "+" (inclusion) and "−" (exclusion)

🅘 *Do you agree? Make sure you understand these steps before continuing.*

We are now ready to derive formulas for merging more than 3 sets. Here is the case for merging 4 sets:

$A \cup B \cup C \cup D$
$= (A + B + C + D)$
$- (A \cap B + A \cap C + A \cap D + B \cap C + B \cap D + C \cap D)$ (4.3)
$+ (A \cap B \cap C + A \cap B \cap D + A \cap C \cap D + B \cap C \cap D)$
$- A \cap B \cap C \cap D$

Chapter 4: Inclusion-Exclusion Principle (Venn Diagram)

You may agree that the 4-set formula is already a bit tedious. It is extremely important not to miss any terms in each block. For example, the second block which contains intersection of any 2 sets should have 6 terms. Why 6? This is because C_4^2, i.e. selecting two sets from 4 possible choices.

4.4 Counting Numbers

Counting is also a popular topic in the number theory. Many questions in math competitions are related to multiples and factors. Often, we will need to employ the *inclusion and exclusion principle* and Venn diagrams to sort out issues related to common multiples.

Example 4.4.1

Among positive integers not exceeding 1000, how many are relatively prime to 1000 itself?

Solution

Let's first count how many are NOT relatively prime to 1000.

Because $1000 = 2^3 \times 5^3$, numbers which are not relatively prime to 1000 must be multiples of 2 or 5. From *Example 4.2.1*, we know the count is 600.

Therefore the count of numbers that are relatively prime to 1000 is $1000 - 600 = 400$.[1]

Done.

---

[1] This is an example of so-called *Euler's ϕ formula* which will be discussed in the book *Number Theory*. Using the ϕ formula, we can directly obtain the result as $1000 \times (1 - \frac{1}{2}) \times (1 - \frac{1}{5}) = 400$.

Chapter 4: Inclusion-Exclusion Principle (Venn Diagram)

4.5 Practice

Practice 1

A group of 30 students participated in after-school sport activities including soccer, basketball, and baseball. Each student participated in at least one activity, but not all three. If 20 signed up for soccer, 12 for basketball, 10 for baseball, 6 for both soccer and basketball, and 2 for both basketball and baseball, how many students signed up for both soccer and baseball?

Practice 2

How many positive, proper fractions are there with a denominator of 2015, in the simplest form?

Practice 3

Find the number of prime numbers that are not exceeding 100.

Chapter 5

Manual Counting

5.1 When to Count Manually

First of all, we should avoid manual counting whenever possible, as it is often quite tedious and prone to errors.

With this said, however, there are two typical scenarios where we would manually count:

1. The problem is simple enough

2. There are no other obviously better approaches

5.2 Systematic Counting

When performing manual counting, it is pivotal to avoid double counting and under counting. How? One tip is to count systematically.

We demonstrate a systematic approach in the following example.

Chapter 5: Manual Counting

Example 5.2.1

Joe wrote down the numbers from 1 to 100. How many times did he write down the figure "2"?

At first glance, a natural method is to list 2, 12, 20, 21, 22, ... It may be workable to count in this way to 100. But if the question asks for 100,000 instead of 100, one may soon find oneself less and less confident about whether any occurrence has been missed.

A more effective process is to count systematically by places:

Solution

We note that a digit (e.g. 2) will appear in:

- the "ones" place: once every 10 numbers
- the "tens" place: 10 times every 100 numbers
- the "hundreds" place: 100 times every 1000 numbers
- ...

Therefore, the answer to the question can be calculated using the following expression:

$$1 \times \frac{100}{10} + 10 \times \frac{100}{100} = 20$$

Done.

If the question asks for 100,000 instead of 100, we can quickly get the answer with a similar expression:

$$1 \times \frac{100000}{10} + 10 \times \frac{100000}{100} + 100 \times \frac{100000}{1000} + \cdots$$

Chapter 5: Manual Counting

Exercise 5.2.1

If you write all the numbers from 1 to 100,000 one by one, how many times will "3" be written?

Exercise 5.2.2

If you write all the numbers from 1 to 500 one by one, how many times will "3" be written?

Exercise 5.2.3

If you write all the numbers from 1 to 500 one by one, how many times will "7" be written?

Caution: what is the difference between Exercise 5.2.2 and Exercise 5.2.3?

5.3 Tree Analysis

Tree analysis, as indicated in its name, employs a systematic decision tree to help list all the possible cases. It proves to be a powerful mechanism in manual counting.

Example 5.3.1

How many different routes are there from A to C without passing any node twice?

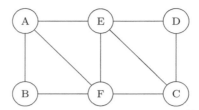

Since it is not easy to identify a direct formula to solve this problem, we turn to the tree analysis to visualize the paths as illustrated below.

Solution

The starting point A is the root node. We then gradually expand the tree by adding more nodes. It is less likely to miss a path with this approach.

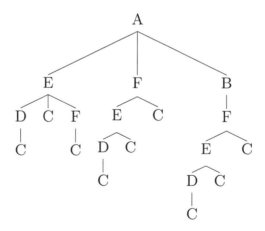

Therefore the answer is 9 routes.

Done.

Chapter 5: Manual Counting

Sometimes it may be useful to simplify the tree. In this example, we note that we can eliminate point B but count two paths from A to F. The simplified tree is shown below:

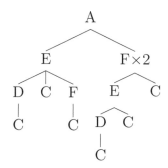

The "×2" part indicates that the count in this branch should be multiplied by 2.

5.4 Hybrid Question

Some manual counting problems are related to other math subjects, such as the number theory. These questions quite often involve multiples and factors.

Example 5.4.1

How many ending "0"s does 99! have?

Clearly it is too laborious to actually complete the multiplication and then count the ending "0"s. We need another approach to find the answer.

First, let's investigate what the phrase "ending '0's" means.

- A number with one ending "0" means it is a multiple of $10^1 = 10$
- A number with two ending "0"s means it is a multiple of

Chapter 5: Manual Counting

$10^2 = 100$

- A number with three ending "0"s means it is a multiple of $10^3 = 1000$

- ...

- A number with n ending "0"s means it is a multiple of 10^n

🔘 *Do you agree?*

Therefore the question is equivalent to finding the largest integer n such that 99! can be divided by 10^n.

To have a factor of 10, we need one "2" and one "5". The question is then equivalent to counting the number of "2"s and "5"s in prime factorization of 99!. Clearly, there are more "2"s than "5"s. Why?

To see this, let's examine a smaller number, such as 5!:

$$5! = 1 \times 2 \times 3 \times 4 \times 5$$
$$= 1 \times 2 \times 3 \times 2^2 \times 5$$
$$= 2^3 \times 3 \times 5$$

This means 5! has three "2"s and only one "5". With the same logic, we only need to count the number of "5"s in 99!, and the result is what we are looking for.

Counting "5"s can be done with the following approach:

- For every 5 consecutive integers, there is a multiple of 5, count once

- For every 25 consecutive integers, there is a multiple of 25, thus count an <u>additional</u> one time

- For every 125 consecutive integers, there is a multiple of 125, thus count an <u>additional</u> time

- ...

The final answer to this question can be calculated as follows

$$\left\lfloor \frac{99}{5} \right\rfloor + \left\lfloor \frac{99}{25} \right\rfloor = 19 + 3 = 22$$

where $\lfloor x \rfloor$ is the *floor* operator which returns the largest integer not exceeding x.

Chapter 5: Manual Counting

5.5 Practice

Practice 1

Find the number of positive integers that satisfy both conditions below:

- Is not exceeding 100
- Is a product of 2 prime numbers

Practice 2

There are two sets {1,3,5,...,23} and {2,4,6,...,24}. If we select one number from each set and add them together, how many different results can we have in total?

Practice 3

There are 3 red balls and 3 green balls in a jar. If we retrieve one ball at a time, and cannot take out more red balls than green balls at any time, how many different ways are there to retrieve these balls?

Practice 4

Positive integers, such as 12321, are called palindromes because they stay the same if the order of their digits is reversed. Find the number of palindromes that are not exceeding 100,000.

Chapter 5: Manual Counting

Practice 5

Count the number of different ways to distribute 6 concert tickets, seating numbers 1 to 6, among 4 students so that:

- every student gets either 1 or 2 tickets, and
- if 2 tickets are given to the same student, these 2 seats must be next to each other

Chapter 5: Manual Counting

Chapter 6

Counting Techniques and Patterns

Now that we have introduced the three counting principles and the two important formulas, this chapter will explore some frequently used techniques and well-studied patterns in solving challenging problems. Your counting skill is highly influenced by your familiarity with these patterns and your ability to apply relevant techniques.

6.1 Count like a Pro

To be an expert in counting, you must be able to speak like a counting pro. This means that you must have the ability to translate a problem from plain English to the proper counting language. Often, if you can describe a problem properly, you are half-way towards working out the final answer.

Take a look at this example:

Chapter 6: Counting Techniques and Patterns

Example 6.1.1

How many line segments can you find in the following diagram?

You can of course count it manually. But can you describe this problem in the proper counting language?

▶ *Pause: think about it before continuing ...*

First, what is the counting language? The most frequently used counting language is a statement which is similar to the following:

> choose k from n choices, when order is [not] important

When *order* is not important, it is combination. Otherwise it is permutation.

Solution

Using the counting language, we can describe this problem as: *select 2 endpoints from 6 choices, when order is not important.*

Why is this description equivalent to the original problem? This is because a line segment is determined by two endpoints. Therefore picking two endpoints is the same as deciding a line segment. The sequence of choosing the two endpoints is obviously irrelevant.

Do you agree?

This counting language implies that the answer is $C_6^2 = 15$.

Done.

Chapter 6: Counting Techniques and Patterns

This is a good example of what it means to be a trained mind from the counting perspective. An intelligent person who is not trained may solve this problem by counting manually in a systematic way. For example, he may notice the following pattern: there are 5 segments of unit length, 4 segments of double length, and so on. Therefore the total count is $5 + 4 + \cdots + 1 = 15$.

The answer is correct and the solution seems reasonable. But this is not what a trained counting mind will do.

Following is another example of reframing a question in the counting language:

Example 6.1.2

How many shortest paths are there from point A to B in the following diagram?

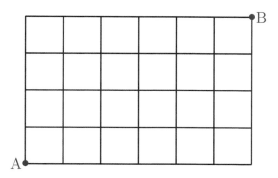

⏸ *Pause: think about it before continuing ...*

Solution

We note that a qualified path from A to B must have 10 steps in total. Among these 10 steps, 4 are northbound (going up) and 6 are eastbound (going right). Different combinations of these northbound and eastbound steps correspond to different paths. For example: *NNNNEEEEEE* represents the path that goes north for 4

Chapter 6: Counting Techniques and Patterns

steps and then goes east for 6 steps.

As such, this problem is equivalent to *selecting* 4 *steps to go north from* 10 *steps in total, and order is not important.*[1] Therefore the answer is $C_{10}^4 = 210$.

Done.

Make sure you fully understand these two examples. We will show an alternative solution to *Example 6.1.2* later.

Remember: a fundamental skill to become a pro is the ability to describe a problem like a pro.

6.2 Complementary Counting (Count the Opposite)

We have introduced this technique before in *Example 2.2.1* on *page 7*. In this section, we will discuss its application in greater details.

The following example appears similar to *Example 6.1.2*. However it has a constraint that demands a new approach.

---

[1] It is the same as *selecting* 6 *steps from* 10 *steps, and order is not important.*

Example 6.2.1

How many shortest paths are there from point A to B **without** passing C in the following diagram?

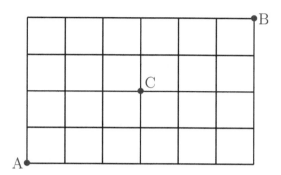

There are different ways in solving this problem. In this section, we will use the complementary counting technique. We will introduce an alternative approach in the next section.

🕚 *Try it yourself before checking the solution.*

Solution

Let's first count how many routes go through point C. This is equivalent to a two-step process: from A to C and then from C to B. Each step is a mini version of *Example 6.1.2*.

Chapter 6: Counting Techniques and Patterns

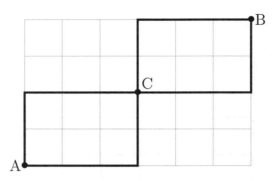

There are $C_5^2 = 10$ different paths from A to C.

There are $C_5^2 = 10$ different paths from C to B.

By the *multiplication principle*, there are a total of $10 \times 10 = 100$ different paths from A to B via C.

We know the number of total paths from A to B is $C_{10}^4 = 210$. Therefore the number of paths from A to B without passing C is $210 - 100 = 110$.

Done.

6.3 Lattice

As indicated in the previous section, *Example 6.2.1* can be solved with another technique called *lattice*.

Study the lattice shown below, and you will notice that each node is marked with a value which equals the total paths that can lead to this node. By the *addition principle*, we can calculate a node's value by adding the values of the nodes immediately below and to its left, if they exist.

As we cannot pass C, we simply black out that point (or mark it as 0).

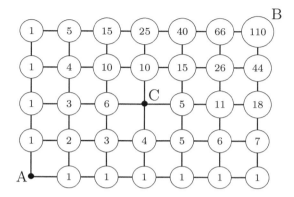

This lattice shows there are 110 paths leading to B without passing C. This result agrees with the answer we have obtained in the preceding section where the complementary counting technique is used.

Lattice is a flexible method, as it can handle multiple "blockages" or irregularly shaped blocks. The algorithm is also simple. Therefore it is particularly suitable for computer programs. In fact, this method is widely used in industries such as finance.

Needless to say, it can also be used to solve *Example 6.1.2* on *page 53* where point C is not involved:

Chapter 6: Counting Techniques and Patterns

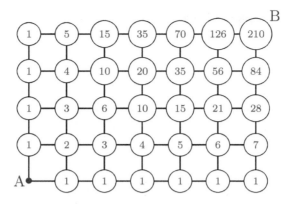

The answer, 210, agrees with our earlier result.

6.4 Bundling

Sometimes certain elements in a problem can be related and regrouped by applying the *bundling* technique, effectively simplifying the original question.

Set forth below is an example to illustrate this technique:

Example 6.4.1

Joe goes to a movie with his parents and his two sisters. The 5 people will sit next to each other on a single row. How many different seating plans do they have? How would you modify the answer if Joe insists on sitting next to his father?

▮▮ *Try it yourself before continuing ...*

Solution

The answer to the 1st question is straightforward: $5! = 120$.

Chapter 6: Counting Techniques and Patterns

To solve the 2nd question, let's imagine that we tie Joe and his father together to create a "superman". This will leave us with just 4 people to arrange freely. There are $4! = 24$ possibilities.

Next, within this "superman", Joe and his father can freely exchange their seats. There are $2! = 2$ possibilities.

By the *multiplication principle*, we find a total of $4! \times 2! = 48$ different seating plans.

Done.

As we can see, the essence of the bundling technique is to satisfy the constraint by relating those elements first (e.g. tie two people together). Doing so will also allow us to decompose a problem with catches into two catch-free sub-problems. Then we can use the *multiplication principle* afterwards to arrive at the final answer.

💡 *Tip: Our objective is to eliminate catches, not to tie people together!*

Let's review another example which can be solved by using the same technique with a tweak.

Example 6.4.2

Fifteen guards and five prisoners stand in a single row. To ensure security, every prisoner must be escorted by two guards, one on each side. The five extra guards can stand anywhere in the row. How many different arrangements can be made?

The catch of this problem is clear. The question is how to eliminate this catch?

⏸ *Pause: think about it before continuing ...*

Chapter 6: Counting Techniques and Patterns

We cannot directly tie a prisoner with two guards, can we? This is because a prisoner can be accompanied by any two guards. Of course we can carefully design a scheme to work it through. But is there a simpler way to solve this problem?

The answer is yes.

Solution

Let's imagine we have 5 three-person chairs and 5 single-person chairs. Every prisoner must sit in the middle of a three-person chair. Guards can sit anywhere except the middle of a three-person chair. This setup can eliminate the catch without loss of generality.

The following diagram shows one qualified arrangement:

$$\underbrace{g}\ \underbrace{g\ P\ g}\ \underbrace{g}\ \underbrace{g}\ \underbrace{g\ P\ g}\ \underbrace{g\ P\ g}\ \underbrace{g\ P\ g}\ \underbrace{g\ P\ g}\ \underbrace{g}\ \underbrace{g}$$

With this setup, we can solve the problem in three steps:

step 1 Setup these 10 chairs in a row. This is equivalent to selecting 5 spots from 10 choices to put the 5 three-person chairs, and order is not important: C_{10}^5.

step 2 Let the 5 prisoners form a line and sit down in sequence: 5!.

step 3 Let the 15 guards form a line and sit down in sequence: 15!.

Therefore the answer is $C_{10}^5 \times 5! \times 15!$.

Done.

What is the difference between this example and *Example 6.4.1*? In the movie seating plan example, we directly tie Joe and his father to form a "superman". In this example, we have created a virtual tie by using a three-person chair. This not only helps us eliminate the catch, but also offers us the freedom to seat anybody later as necessary.

6.5 Circle / Round Table

Circle is a variation of regular counting. The difference is whether rotation is allowed.

Example 6.5.1

How many different seating plans are there for 7 people to sit at a round table? Two seating plans will be considered the same if they are different just by rotating these people.

The catch is obviously the rotation. How to eliminate this catch?

🅘 *Pause: think about it before continuing ...*

There are several approaches in solving this problem. We will discuss two of them.

Solution 1 - Pin an Anchor

Let's pick a person and *pin* him to a fixed seat as the anchor. The remaining 6 people sit freely. These seating plans will not coincide by just rotating these people. Therefore the answer is $6! = 720$.

Done.

Solution 2

If we allow everyone to sit freely without consideration of the rotation restriction, there are altogether $7!$ possibilities.

Chapter 6: Counting Techniques and Patterns

We note that every seating plan can be mapped to 6 other plans by just rotating. This means these 7 seating plans can only be counted as 1 per the rule.

Therefore the answer is $\frac{7!}{7} = 6!$, which agrees with the result above.

Done.

Tip: Does Solution 2 look similar? It is somewhat similar to the derivation of the combination formula from the permutation formula discussed in Chapter 3. Both divide the count of duplicates.

Now we can draw the conclusion from *Example 6.5.1* in the following:

> For n people to sit at a round table, there are $(n-1)!$ different seating plans when rotations count as duplicates.

In other words, the impact of rotation effectively excludes one element from the free seating option.

 Caution: A table cannot flip. Otherwise we should take into consideration the effect of flipping.

Example 6.5.2

Mary plans to make a ring out of 7 beads that are colored differently. How many different styles can she create?

A ring can be rotated and flipped. Therefore the answer is $\frac{6!}{2} = 360$.

Chapter 6: Counting Techniques and Patterns

6.6 Permutation of Combination

This is a hybrid of permutation and combination.

Example 6.6.1

How many different encoded messages can be created with 9 flags that are lined up? Among these flags, 2 are red, 3 are blue, and 4 are green.

Flags in this question can be changed to different letters, or balls with different colors. But they are all asking the same type of questions. Can you work out the solution?

Solution

This problem can be solved in 3 steps:

step 1: Pick 2 red flags out of 9: C_9^2

step 2: Pick 3 blue flags out of the remaining 7: C_7^3

step 3: Pick 4 green flags out of the remaining 4: C_4^4

Therefore the final answer is: $C_9^2 C_7^3 C_4^4 = 1260$.

Done.

How about selecting the blue flags first, and the red color next? We will get $C_9^3 C_6^2 C_4^4$, which leads to the same result. In fact, if you expand either formula, each leads to

$$\frac{9!}{2!3!4!}$$

We can think of such question as an extension of a regular permutation problem. Consider the following: how many different messages can be encoded by using 9 flags with different colors? The answer is simply 9!.

Chapter 6: Counting Techniques and Patterns

Now let's make two of these flags blue. This makes switching these two flags indistinguishable. As a result, the answer will become $\frac{9!}{2!}$. If we make 3 flags blue, the answer will be $\frac{9!}{3!}$. Now let's make 2 other flags red. Along the same logic, we have $\frac{9!}{3!2!}$. We can repeat this process by making the remaining 4 flags green, thus $\frac{9!}{2!3!4!}$.

? *Quiz: What conclusion or pattern have you noticed?*

Assume n objects are divided into k **distinguishable** groups. If group i contains r_i **indistinguishable** objects, then the number of permutations possible can be found by

$$\frac{n!}{r_1! r_2! \cdots r_k!}$$

where $r_1 + r_2 + \cdots + r_k = n$

In our example, we are dividing 9 flags into 3 distinguishable groups with 2, 3, and 4 flags, respectively, in each group. Therefore the answer is $\frac{9!}{2!3!4!}$.

The word *distinguishable* is important. Three flag colors are involved, and hence the three distinguishable groups of flags. Within each group, the objects are indistinguishable. Therefore this type of problems exhibits a permutation pattern with a combination problem at the sub-level.

💡 *Tip: As a learner, you do not need to remember this formula. It is much more important to understand how to solve such problems by using the 3-step logic described in the solution. This will help you avoid making mistakes. Blindly applying formulas without a thorough understanding is a sure recipe of getting tricked.*

6.7 Combination of Combination

You can guess this section is related to the preceding topic.

Example 6.7.1

How many different possibilities are there to divide 6 different books into 3 piles equally?

Solution

This problem can be solved in the following 4 steps:

step 1: Select 2 books to form the 1st pile

step 2: Select 2 from the remaining books to form the 2nd pile

step 3: Select 2 from the remaining books to form the 3rd pile

step 4: Apply the multiplication principle, and divide the result by 3!

Therefore the answer is $\frac{C_6^2 C_4^2 C_2^2}{3!} = 15$

Done.

The first 3 steps are easy to understand. But why is the 4th step necessary? This is because the 3 piles themselves are *indistinguishable*. However the first 3 steps implicitly forced an order of these 3 piles. Therefore we need to convert *permutation* to *combination* by dividing the count of duplicates.

A bit confused? Let's illustrate with a simplified version: *dividing 4 books into 2 piles*. In this case, we only need 2 steps. Assuming A, B, C, and D are the 4 different books, we have:

Chapter 6: Counting Techniques and Patterns

step 1	step 2
AB	CD
AC	BD
AD	BC
BC	AD
BD	AC
CD	AB

There are exactly $C_4^2 C_2^2 = 6$ rows by following the first two steps. However, we spot that the 1st and the 6th rows are the same. So are the 2nd and the 5th rows, and the 3rd and the 4th rows. Why are half of the rows duplicated? The duplication is caused by the order in creating these piles.

Therefore the final answer is $\frac{C_4^2 C_2^2}{2!} = 3$.

So what is the relationship between this example and *Example 6.6.1*? The first 3 steps are exactly the same. In the previous example, the groups are distinguishable by colors. In this example, the piles are indistinguishable. Therefore we arrive at the following conclusion:

> Assume n **distinguishable** objects are divided into k **indistinguishable** groups. If group i contains r_i objects, then the number of arrangements possible can be found by
>
> $$\frac{n!}{r_1! r_2! \cdots r_k!} \cdot \frac{1}{k!}$$
>
> where $r_1 + r_2 + \cdots + r_k = n$

Using this formula, the answer to our example of dividing 6 books into 3 piles equally is $\frac{6!}{2!2!2!} \times \frac{1}{3!} = 15$.

Again, it is more important to remember how to get the solution, not just the formula.

Chapter 6: Counting Techniques and Patterns

6.8 Different Balls in Different Boxes

While appearing similar to the two examples discussed above, this new example is in fact completely different in nature.

Example 6.8.1

Find the number of possible ways to put 6 different balls into 3 different boxes so that each box contains at least 1 ball.

🅿 *Pause: think about it before continuing ...*

Solution

It turns out to be a Venn diagram problem!

Without restriction, there are $3^6 = 729$ ways because each ball has 3 choices. We then apply the complementary counting technique demonstrated in *Section 6.2* to figure out the number of times when at least one box is empty. The difference is the final answer that we are looking for.

Let E_i ($i = 1, 2, 3$) be the set where the ith box is empty.

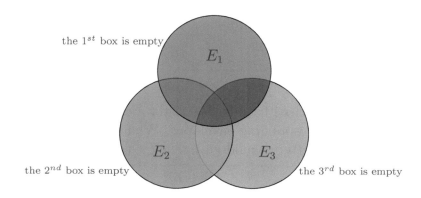

- $|E_1| = |E_2| = |E_3| = 2^6 = 64$

Chapter 6: Counting Techniques and Patterns

- $|E_1 \cap E_2| = |E_2 \cap E_3| = |E_3 \cap E_1| = 1^6 = 1$
- $|E_1 \cap E_2 \cap E_3| = 0$
- $|E_1 \cup E_2 \cup E_3| = 64 \times 3 - 1 \times 3 + 0 = 189$

Therefore the answer is $729 - 189 = 540$.

Done.

We conclude in the following:

> In general, to put m different balls into n different boxes ($m \geq n$):
> - if empty boxes are allowed, there are n^m different ways
> - otherwise there are $n^m - (n-1)^m C_n^1 + (n-2)^m C_n^2 \cdots + (-1)^{n-1} \times 1^m C_n^{n-1}$ ways

6.9 Derangement

Here is another type of problems that can be solved using the inclusion-exclusion principle.

Example 6.9.1

Joe wrote 4 letters to 4 friends. His sister messed up all the letters and their corresponding envelopes. What is the probability that none of the letters would go to their intended recipients?

Solution

There are altogether 4! ways to put 4 letters into 4 envelopes. We therefore just need to count those situations where at least one

Chapter 6: Counting Techniques and Patterns

letter matches its envelope. This can be done using the inclusion-exclusion principle.

Let M_i, $1 \leq i \leq 4$, be the case when the ith letter matches its envelope. Clearly we have:

- $|M_1| = \cdots = (4-1)!$
- $|M_1 \cap M_2| = \cdots = (4-2)!$
- $|M_1 \cap M_2 \cap M_3| = \cdots = (4-3)!$
- $|M_1 \cap M_2 \cap M_3 \cap M_4| = (4-4)!$

It follows that

$$|M_1 \cup M_2 \cup M_3 \cup M_4|$$
$$= C_4^1 \times 3! - C_4^2 \times (4-2)! + C_4^3 \times (4-3)! - C_4^4 \times (4-4)!$$
$$= \frac{4!}{1!} - \frac{4!}{2!} + \frac{4!}{3!} - \frac{4!}{4!}$$
$$= 4! \times (\frac{1}{1!} - \frac{1}{2!} + \frac{1}{3!} - \frac{1}{4!})$$

The count of no-match at all is:

$$4! - 4! \times (\frac{1}{1!} - \frac{1}{2!} + \frac{1}{3!} - \frac{1}{4!})$$
$$= 4! \times (1 - \frac{1}{1!} + \frac{1}{2!} - \frac{1}{3!} + \frac{1}{4!})$$
$$= 4! \times (\frac{1}{2!} - \frac{1}{3!} + \frac{1}{4!})$$
$$= 9$$

Therefore the probability is $\frac{9}{4!} = \frac{3}{8}$.

Done.

Such problems are called derangement. The total count is usually written as D_n. In this example, it is D_4. It may come in handy if you familiarize yourself with the formulas below:

> **Derangement**
>
> $$D_n = n!(1 - \frac{1}{1!} + \frac{1}{2!} - \frac{1}{3!} + \cdots + (-1)^n \frac{1}{n!})$$
>
> Correspondingly, its probability is
>
> $$\frac{D_n}{n!} = 1 - \frac{1}{1!} + \frac{1}{2!} - \frac{1}{3!} + \cdots + (-1)^n \frac{1}{n!}$$

It can be shown that, as n increases to infinity, this probability approaches $e^{-1} \approx 0.3679$, where e is the famous math constant.

6.10 Mapping

Mapping is another practical method when it is difficult to count directly.

Example 6.10.1

On a $m \times n$ chessboard, how many different L-shaped blocks (as shown below) can be formed?

For example, on a 2×2 chessboard, 4 such L-shaped blocks can be formed.

🔘 *Pause: think about it before continuing ...*

Since it does not appear to be easy to count directly, we will map it out.

The clue comes from the center point of the L-shaped. For an L-shaped block to be on the chessboard, its center point cannot be on a border. On the other hand, every inner point can be the center of four L-shaped blocks.

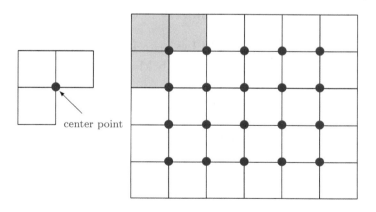

There are altogether $(m-1)(n-1)$ inner points. And for each center point, there are 4 L-shaped blocks. Therefore a total of $4(m-1)(n-1)$ L-shaped blocks can be formed.

6.11 Symmetry

Symmetry is not as elementary as it seems. In counting, to be symmetrical means the two parts have the same count, even if we may not know the exact value.

Example 6.11.1

Seven people, including Joe and Mary, form a line:

(a) How many different ways can they line up?

(b) How would your answer change if Joe must stand ahead of Mary but not necessarily next to her?

The answer to the first question is: $P_7^7 = 7! = 5040$.

How about the second question? The key is to eliminate the catch that Joe must stand ahead of Mary. There are different ways to eliminate this catch, but symmetry is the easiest technique.

The arrangements in which Joe stands ahead of Mary is symmetrical to the arrangements in which Mary stands ahead of Joe.

This statement is correct because we can switch the positions of Joe and Mary, but nobody else, to create a pair of corresponding arrangements. One of this pair has Joe standing ahead of Mary while the other has Mary standing ahead of Joe.

Chapter 6: Counting Techniques and Patterns

Any arrangements will have either Joe standing ahead of Mary, or vice versa. The symmetry argument implies that the two cases have equal counts. Therefore the answer is half of the total possibilities, i.e. $\frac{1}{2}P_7^7 = 2520$.

> Symmetry means *it makes no difference when switching the positions of the two parties*. As indicated in this example, there is no difference in the number of possibilities whether Mary stands ahead of Joe, or vice versa. Such exchangeability often offers a crucial hint at symmetry.

Let's examine a more challenging example, in which the symmetry technique offers an easy solution.

Example 6.11.2

Joe and Mary flip a coin $(n+1)$ and n times, respectively. What is the probability that Joe gets more heads than Mary does?

At first glance, there seems to be no symmetry here because $(n+1)$ and n apparently are not equal. But think twice as the following diagram illustrates.

Chapter 6: Counting Techniques and Patterns

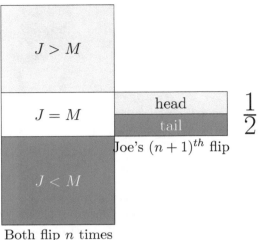

Both flip n times

First, let's analyze their first n flips. There are three possible cases: Joe gets more heads; Joe gets the same number of heads as Mary does; and Joe gets fewer heads. We note that the 1st and the 3rd cases have the same possibility, as there is in fact no difference between Joe's n tosses and Mary's. Both have the same chance to get more heads.

In the 1st case, if Joe has already gotten more heads in the first n tosses, he will always get more heads in this game regardless of the outcome of his $(n+1)^{th}$ flip. Therefore this case will be qualified in the final answer.

In the 3rd case, if Joe gets fewer heads after having flipped the coin n times, he can at most get the same number of heads as Mary by flipping one more time. However the question asks for "more" heads. Therefore this case will be disqualified in the answer regardless of his $(n+1)^{th}$ attempt.

In the 2rd case where Joe and Mary have tied, the final outcome depends on Joe's $(n+1)^{th}$ toss. Clearly, it is a 50-50 chance for Joe to flip a head up in his last try. Therefore there is a 50% chance in this case that Joe will get more heads than Mary will.

As such, we can see the top part (where Joe gets more heads) and the bottom part (where Joe gets fewer heads) in the diagram are symmetrical. Therefore the answer to the original question is $\frac{1}{2}$.

6.12 Cut the Rope

Let's work through a problem first:

Example 6.12.1

Roll a standard 6-sided die three times. What is the probability that their sum is 8?

While it is easy to calculate the total possibilities as 6^3, it is not as straightforward to count the scenarios where the sum is 8.

Fortunately, such problems belong to a well-studied pattern, for which a well-represented solution is available.

Another similar and popular question in this group of pattern is to count the total **positive integer** solutions of equation $x_1 + x_2 + x_3 = 8$.

Do you see these two problems are the same?

The wording and the numbers here are carefully selected. By specifying *positive integers* and limiting the sum of 3 numbers at 8, we have guaranteed that none of the numbers will exceed 6, which is the largest number we can get on a standard die.

If we replace 8 with 9, then the two questions are no longer equivalent, because (7,1,1) is a solution to the equation, but does not apply in the die-rolling case.

Chapter 6: Counting Techniques and Patterns

Likewise, if we relax the word *positive* with *non-negative* in the equation problem, these two questions are no longer the same, as zero does not appear on any side of a standard die.

These two variances will be discussed in further detail in the following two sections. Here we will focus on *Example 6.12.1*.

Solution

Suppose we have 8 balls tied to a rope as shown. Then the problem is equivalent to cutting the rope into 3 segments, each of which is corresponding to the result of one roll. For example, the diagram shown below means the 1st roll shows a 2, and the 2nd and the 3rd rolls both yield a 3.

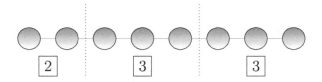

If we use the *mapping* language as discussed in *Section 6.10* on *page 70*, there is a one-to-one mapping between the outcomes of die rolling and those of rope cutting. Therefore the problem can be described as cutting the rope twice into 3 segments along the 7 intervals between these balls. In the counting language, it means *selecting 2 from 7 intervals to make the cuts, and order is not important*. This give us the answer of $C_7^2 = 21$.

Done.

Make sure you fully understand this example before continuing. The next two sections are built upon this example.

6.13 Knives and Balls

As mentioned earlier, the preceding example is equivalent to counting **positive** integer solutions of $x_1 + x_2 + x_3 = 8$. What if **positive** is replaced with **non-negative**?

Example 6.13.1

How many **non-negative** integer solutions does the equation $x_1 + x_2 + x_3 = 8$ have?

⏸ *Pause: think about it before continuing ...*

Solution

We will still illustrate with 8 balls. However, instead of tying the balls together with a rope, we arrange them together with two knives this time. The number of balls before the 1st knife, between the two knives, and after the 2nd knife represents the value of x_1, x_2, and x_3, respectively.

For example,

This arrangement is corresponding to $2 + 2 + 4 = 8$. A cross represents a knife.

This arrangement is corresponding to $0 + 1 + 7 = 8$.

This arrangement is corresponding to $8 + 0 + 0 = 8$.

Chapter 6: Counting Techniques and Patterns

Therefore the solution is equivalent to *selecting 2 from a total of 8 + 2 = 10 spots to place these two knives, and order is not important*. Therefore, the answer is $C_{10}^2 = 45$.

Done.

💡 *Tip: A formula for this pattern can be generalized as C_{n+r-1}^{r-1}, where n is the sum (i.e. 8 in this example) and r is number of variables (i.e. 3 in this example).*

Again, you do not need to memorize one more formula. If you understand and can remember the solution instead, it would benefit in the long run and help mitigate the risks of being tricked by subtle differences in problems.

6.14 Generating Function

Now let's examine the second variation of *Example 6.12.1*.

Example 6.14.1

Roll a standard 6-sided die three times. What is the probability that their sum is 9?

It turns out to be a completely different question from *Example 6.12.1*, even though the difference is simply replacing the sum of 8 with 9. This is because the largest number on the sides of a die is 6, therefore a combination of 7, 1, and 1 cannot be a solution.

With this seemingly minor difference, the complexity of the problem increases significantly.

Chapter 6: Counting Techniques and Patterns

In this case, the gap is only 1, so the problem is manageable by employing the *complementary counting* technique. Afterwards, we will introduce a powerful method to solve such problems.

Solution 1 (Complementary Counting)

Without constraints (i.e. allow the die to produce a 7), there are $C_8^2 = 28$ possibilities by applying the *Cut the Rope* technique.

The only illegitimate cases included in the above count are those when the die settles on a 7. As there are 3 rolls, there are 3 illegitimate cases.

Therefore the total qualified cases are: $28 - 3 = 25$. The probability is $\frac{25}{6^3} = \frac{25}{216}$.

Done.

The complementary counting technique works, but it requires a little bit maneuver. For example, if the sum is 10, we will have to apply the *principle of inclusion and exclusion* or other techniques.

Now it is the time to introduce a very powerful method: *generating function*. Similar to the *lattice* method, the concept of *generating function* is intuitive, and the algorithm is straightforward. However it may be tedious to calculate manually. Therefore it is best to be implemented by computer software.

Solution 2 (Generating Function)

It is claimed that the count of the total possibilities is the coefficient of x^9 in the following polynomial after expansion:[2]

$$(x + x^2 + \cdots + x^6)(x + x^2 + \cdots + x^6)(x + x^2 + \cdots + x^6)$$

---

[2] If you do not want to expand manually, visit http://www.wolframalpha.com/ and then type *expand* $(x + x^2 \cdots)$. You will have to type the polynomial in full and then look for the coefficient of x^9.

The coefficient is 25, and therefore the probability is $\frac{25}{216}$.

Done.

Do you understand and agree with the claim?

🅿 *Pause: think about it before continuing ...*

To see why the claim is correct, we note that any term in the expanded form must be a product of three terms. Each of these three terms comes from one of the three respective brackets. The exponent of the final term is the sum of the exponents of these three terms.

For example, a solution of $2 + 3 + 4 = 9$ is equivalent to choosing x^2, x^3, and x^4 from these three brackets, respectively. The total count is thus the coefficient of the x^9.

Because we have restricted every bracket to x^1 to x^6, each term can only contribute a number from 1 to 6, corresponding to the numbers on the sides of the die.

The flexibility of this method is that we can simply adjust each bracket to accommodate different constraints. For example, if our dice is not standard and is painted with 0 to 5, instead of 1 to 6, we can simply adjust each bracket to $1 + x + \cdots + x^5$.

6.15 Modeling

It is debatable whether modeling is a separate technique. For instance, the mapping technique discussed in *Example 6.10.1* on *page 71* can be viewed as an example of modeling too.

Regardless, modeling is worthy of emphasis on its own. Our primary objective is to hone counting skills. If we come across

Chapter 6: Counting Techniques and Patterns

a challenging problem where a solution is not apparent, we may consider modeling so it becomes easier to solve.

Let's consider an example:

Example 6.15.1

Two schools hold an annual chess tournament. Each team has 10 participants with a pre-determined order to play. In each round, the loser is eliminated and the winner plays against the next player from the opponent team. When all the 10 players of one team are eliminated, the contest finishes.

If we record the progress of the entire tournament (i.e. who plays against whom, and in which round), how many different outcomes are possible?

This problem can be solved with casework. However such an approach is obviously tedious. Modeling comes in handy in tackling this type of problems.

Solution

The problem is equivalent to forming a line using two sets of balls, each labeled from 1 to 10. Each set represents a team. Within each set, the 10 balls are always in order. Two sets are merged according to the rule that the defeated is always put before the winner.

This problem thus becomes *selecting* 10 *from a total of* 20 *spots, and order is not important*. These 10 spots will belong to one team and the rest belong to the other.

Therefore the answer is C_{20}^{10}.

Done.

Chapter 6: Counting Techniques and Patterns

We present a simplified diagram to explain this modeling process.

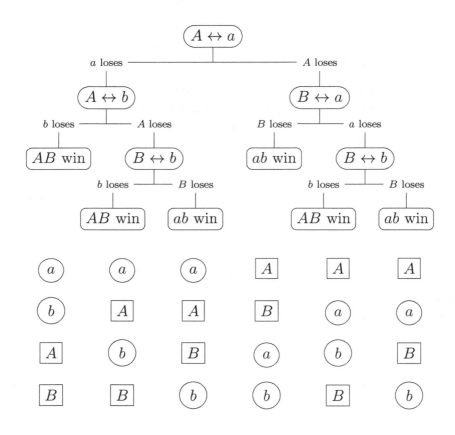

This above diagram details all possible outcomes, if each team only has two players. One team consists of players A and B, while the other is represented by a and b.

The tournament progress in the upper half of the diagram can be translated to the outcome arrangement in the lower half of the chart. Simply record the defeated along the path and then append the winner and his fellow team member who does not need to play, if applicable. Each arrangement should correspond to one outcome. For instance, the first combination, $a - b - A - B$, means both a and b are defeated by A since both are positioned immediately in front of *A*. *B* does not have to play in this outcome.

While it may not appear to be straightforward at first glance, modeling helps the brain process through some challenging and complex problems, and therefore remains a compelling and robust technique.

6.16 Shortcut

Some problems may be solved by using shortcuts without "real" counting. Let's see an example.

Example 6.16.1

In a single elimination tournament, all teams will be grouped into pairs to compete. The loser of each pair will be eliminated. Winners will be regrouped into new pairs and compete again. This process will be repeated until a champion is determined.

How many matches will be played in a 32-team tournament?

Conventional Solution

We note that there will be 5 rounds of matches.

Round	1	2	3	4	5
Matches	16	8	4	2	1

Therefore a total of $16 + 8 + 4 + 2 + 1 = 31$ matches will be played.

Done.

Chapter 6: Counting Techniques and Patterns

This solution works well in this case where the number of teams can be written as 2^n, specifically $32 = 2^5$. If this condition does not hold (e.g. the number of teams is 31), we will need to make assumption on how to deal with that "lone" team in each round.

Alternative Solution

We note that one match will eliminate one team. In order to determine the champion, we need to eliminate 31 teams. Therefore 31 matches are required.

Done.

6.17 Practice

Practice 1

Seven people form a line. If A must stand next to B, and C must stand next to D, how many possibilities are there?

Practice 2

How many triangles are there in this diagram?

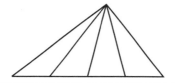

Practice 3

How many rectangles are there in the following diagram?

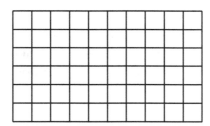

Chapter 6: Counting Techniques and Patterns

Practice 4

Team MAS won a total of 10 gold medals in a 6-day tournament. It won at least one gold medal every day. How many different possibilities are there to count the number of gold medals won each day?

Practice 5

Find the number of positive integer solutions to the following equation:
$$x_1 + x_2 + \cdots + x_6 = 20$$

Practice 6

Find the number of non-negative integer solutions to the following equation:
$$x_1 + x_2 + \cdots + x_6 = 20$$

Practice 7

Joe goes to a supermarket to buy 10 cakes. There are 6 different types of cakes, and each type has a sufficient quantity. How many different combinations of cakes can Joe have?

Practice 8

If integers x, y, and z satisfy $0 \leq x \leq 2$, $1 \leq y \leq 3$, and $1 \leq z \leq 2$, how many different combinations of (x, y, z) can satisfy $x + y + z = 4$?

Chapter 6: Counting Techniques and Patterns

Practice 9

A jar contains 5 red balls, 4 green balls, and 3 blue balls. If we take out one ball at a time and record its color, how many different possibilities can we have after all the balls are taken out and their colors are recorded?

Practice 10

Seven people with different last names sit at a round table. What is the probability of them sitting in the alphabetical order of their last names either ascendingly or descendingly?

Practice 11

Let integers $1 \leq a_1 < a_2 < a_3 < a_4 < a_5 \leq 25$. The difference between any 2 of these 5 integers is not less than 4. How many different choices of the integers are there?

Practice 12

To arrange 3 white balls and 8 black balls in a circle so that every two white balls must be separated by at least two black balls. How many different arrangements do we have? (Two arrangements will be considered the same if they are different just by rotating the balls.)

Practice 13

Find the number of different ways to assign 5 different tasks to 4 workers so that every worker gets at least one task.

Chapter 6: Counting Techniques and Patterns

Practice 14

Four people attending a gathering bring a gift each for exchange. Everybody shall get one gift that is not his own. How many different ways are there to exchange these gifts?

Chapter 7

Elementary Probability

A fundamental basis for some of the counting patterns we have covered thus far lies in the topic of probability. Probability is an advanced and complex subject. This chapter, however, will focus only on elementary probability that is sufficient for middle school and high school students.

7.1 Probability Defined

The concept of probability is intuitive, and we have been using it and calculating it in previous chapters. A more formal definition is as follows:

> **Definition 7.1.1 Probability**
>
> Probability is a measure of the likelihood that an event will happen. An event's probability must be between 0 and 1, inclusive.

Chapter 7: Elementary Probability

A probability of 0 means the event is impossible, while a probability of 1 means the event will certainly happen. The closer a probability is to 0, the more improbable it is that the event will happen. The closer a probability is to 1, the more certain we are that the event will happen.

The classical viewpoint of probability assumes that the events involved are countable, and therefore, the probability of an event happening is computed as follows:

$$\text{probability} = \frac{\text{number of the qualified events}}{\text{number of the total events}}$$

Often, some events will occur with equal probability. In such cases, we can obtain the probability of an individual event happening by dividing 1 by the total number of possibilities. For example:

- The probability of a coin to flip head (or tail) is 50%

- The probability of a standard 6-sided die to roll any one side is $\frac{1}{6}$

7.2 Independent and Exclusive Events

If you review the probability related examples discussed in the previous chapters, you may note that all our computations are based on the definition presented in the preceding section..

To understand what this statement means, consider a simple example:

Example 7.2.1

Roll a standard 6-sided die, what is the probability that its reading is an even number?

Chapter 7: Elementary Probability

The solution is: $\frac{3}{6} = \frac{1}{2}$ or 50%.

However, a person skilled in probability may calculate as below:
$$\frac{1}{6} + \frac{1}{6} + \frac{1}{6} = \frac{1}{2} = 50\%$$

❚❚ *Do you see the difference between these two approaches?*

Mathematically, they are the same. But conceptually, they are different.

- The 1st approach counts the total possibilities of the die showing 2, 4, and 6, and treats them as a single combined event.

- The 2nd approach treats them as three independent events, and calculates each probability separately before aggregating them.

This difference appears to be subtle, but it is important. While the 1st approach may be what you are more comfortable with at first, you should start to use the 2nd approach when working on probability related questions. This is because the latter is more flexible, especially when you are solving more complex probability problems. You will see examples that illustrate this point later in this chapter.

To be able to apply the 2nd approach effectively, we need to understand a few important concepts.

Definition 7.2.1   Independent Events

If the outcome of an event has no effect on the probability of the outcome of another event, then these two events are independent.

To state it another way, two events are considered independent if the outcome of the 2nd event does not depend on the outcome of the 1st event.

Chapter 7: Elementary Probability

For example:

- If we roll a die twice, the outcomes of these two rolls are independent events.

- From a jar containing balls with mixed colors, draw two balls one at a time, without replacement. Whether a red ball is drawn in the 1st try impacts the outcome of the 2nd draw. Therefore, these two draws are not independent events. However, if the 1st ball is placed back in the jar after being drawn, then these two draws become independent events.

> **Definition 7.2.2   Mutually Exclusive Events**
>
> If two events can not occur at the same time, they are mutually exclusive.

For example:

- If we roll a die, the event of getting a 1 and the event of getting a 2 are mutually exclusive.

- If we draw a poker card, the event of getting an Ace and the event of getting a heart are NOT mutually exclusive.

When two or more events overlap, such as drawing a card that is an Ace and drawing a card that is a heart, the probability involving these events is called a joint probability. The general rules of calculating joint probabilities follow.

> **Property 7.2.1   Joint Probability Calculation (I)**
>
> Let P_A be the probability that event A will occur and P_B be the probability that event B will occur.
>
> - If A and B are independent, the probability of both A and B occurring is $P_A \cdot P_B$.
>
> - If A and B are mutually exclusive, the probability of either A or B occurring is $P_A + P_B$.

Chapter 7: Elementary Probability

💡 *Tip: Do you see similaritites between these two rules and the situations where we have applied the multiplication principle and the addition principle?*

In the die-rolling example on page 90, the expression $\frac{1}{6} + \frac{1}{6} + \frac{1}{6}$ uses the property of mutually exclusive joint probability.

Let's review another example:

Example 7.2.2

Roll a die three times, what is the probability of

(a) getting a 2 three times?

(b) getting an even number three times?

Solution

(a) Whether the 1st roll yields a 2 does not impact the outcome of the 2nd and the 3rd rolls. Therefore, the three rolls are independent events. The answer is

$$\frac{1}{6} \times \frac{1}{6} \times \frac{1}{6} = \frac{1}{216}$$

(b) Each independent event (e.g. each roll) contains three mutually exclusive events (i.e. 2, 4, 6). Therefore the answer is

$$\left(\frac{1}{6} + \frac{1}{6} + \frac{1}{6}\right) \times \left(\frac{1}{6} + \frac{1}{6} + \frac{1}{6}\right) \times \left(\frac{1}{6} + \frac{1}{6} + \frac{1}{6}\right) = \frac{1}{8}$$

Done.

When studying probability, we often use the concept and notation of *set*. If A and B are two events,

- the event of either A or B occurring is written as $A \cup B$

Chapter 7: Elementary Probability

- the event of both A and B occurring is written as $A \cap B$

Using this notation, we can extend the *principle of inclusion and exclusion* to help us calculate joint probability, regardless of whether the underlying events are mutually exclusive.

If we use $P(X)$ to denote the probability that event X will occur, we will have

> **Property 7.2.2 Joint Probability Calculation (II)**
>
> $$P(A \cup B) = P(A) + P(B) - P(A \cap B) \qquad (7.1)$$

Does this formula look familiar?

When A and B are mutually exclusive, they will not occur at the same time. In this case $P(A \cap B) = 0$. As a result, *Equation 7.1* becomes $P(A \cup B) = P(A) + P(B)$.

Make sure you understand this equation. The *inclusion-exclusion principle* appears frequently in probability calculation.

Let's revisit *Example 2.2.1* on *page 7*. We have presented three alternative counting solutions to this problem in *Chapter 2*. Now we will solve it with the probability calculation formula discussed above.

Randomly draw a card from 1 to 10 twice, with replacement (i.e. the card is put back after each draw). What is the probability that the product of these two cards is a multiple of 7?

Solution

Let A be the outcome of getting a 7 in the 1st draw and B be the outcome of obtaining the same card in the 2nd draw. Obviously these two outcomes are independent and are not mutually exclusive. As $P(A) = P(B) = \frac{1}{10}$, we have

$$P(A \cap B) = \frac{1}{10} \times \frac{1}{10} = \frac{1}{100}$$

The event that the product is a multiple of 7 is equivalent to either A or B occurs, i.e. $P(A \cup B)$:

$$P(A \cup B) = P(A) + P(B) - P(A \cap B) = \frac{1}{10} + \frac{1}{10} - \frac{1}{100} = \frac{19}{100}$$

Done.

Do you see the difference between this solution and the three solutions discussed in Chapter 2?

7.3 Probability Tree Diagram

Similar to the decision tree explained in *Section 5.3* on *page 43*, a probability tree diagram assigns a probability to each branch of a tree structure to help with computation.

Let's illustrate this technique using an example.

Chapter 7: Elementary Probability

Example 7.3.1

A bug crawls on a grid from A to L as shown below. In order to get to L as quickly as possible, it only moves north or east on the diagram. At each point, the possible directions are assigned an equal probability. For example, at point B, it has a 50% chance of moving to E and a 50% chance of moving to C. However at C it is certainly moving towards F.

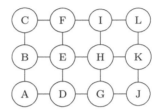

What is the probability that the bug will pass the following points en route?

(i) B (ii) C (iii) E (iv) F

(i) The probability of passing B is: $50\% = \frac{1}{2}$.

(ii) The probability of passing C may be obtained as $\frac{1}{2} \times \frac{1}{2} = \frac{1}{4} = 25\%$. To reach C, the bug must travel from A to B, and then from B to C. At both A and B, it has a 50% chance to go north. As moving from A to B and crawling from B to C are independent events, the probability is their product, i.e. $\frac{1}{2} \times \frac{1}{2}$.

We have constructed a probability tree diagram, as shown below, to describe it more intuitively:

Chapter 7: Elementary Probability

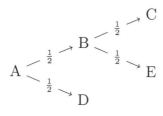

It suggests that at A, the bug has a 50-50 chance of crawling to either B or D. Likewise, at B, the bug can proceed to either C or E, each with a 50% chance.

In a probability tree diagram:

- Each node (e.g. B and D) represents a mutually exclusive event.

- Each path represents a series of independent events.

- Fractions, decimals, or percentages shown along the paths are the corresponding probabilities.

To calculate the probability of reaching any node, we can multiply all the probabilities leading to that node. If there are multiple ways leading to that node, we shall add up all those multiplication results.

It is also unnecessary to draw all the possible branches. Instead, we can focus only on those that will lead to the target node. For example, the tree above can be simplified to the one below:

$$A - \tfrac{1}{2} \to B - \tfrac{1}{2} \to C$$

(iii) The probability of passing E can be easily calculated by using this method.

Chapter 7: Elementary Probability

```
        ┌ 1/2 → B — 1/2 → E
    A ──┤
        └ 1/2 → D — 1/2 → E
```

The answer is $\frac{1}{2} \times \frac{1}{2} + \frac{1}{2} \times \frac{1}{2} = \frac{1}{2}$.

(iv) The probability of passing F can be calculated by using the following diagram:

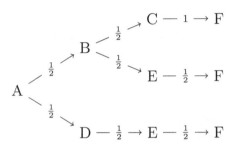

Therefore the probability of passing F is

$$\frac{1}{2} \times \frac{1}{2} \times 1 + \frac{1}{2} \times \frac{1}{2} \times \frac{1}{2} + \frac{1}{2} \times \frac{1}{2} \times \frac{1}{2} = \frac{1}{2}$$

Exercise 7.3.1

Calculate the probabilities of passing H and J, respectively.

Exercise 7.3.2

What value will we get if we add the probabilities of passing F, H, and J? Explain the reason.

Chapter 7: Elementary Probability

7.4 Geometric Probability

Geometric probability can be a complex subject, but our discussion in this section will not involve advanced math such as calculus.

Let's start with a simple example.

Example 7.4.1

Randomly select a point inside a circle. What is the probability that the distance between the point and the center of the circle does not exceed half of the radius?

The answer is: $\dfrac{\pi(\frac{1}{2}r)^2}{\pi r^2} = \dfrac{1}{4}$

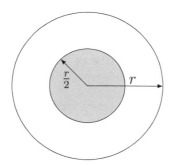

There is a fundamental difference between this example and all the other examples we have discussed so far.

Prior to *Example 7.4.1*, the number of qualified cases and the number of total cases are both countable. For example, when a die is rolled, the number of possible cases totals 6.

However, in this example, the number of possible cases is the number of total points inside a circle, which is infinite! Same is true for the number of total qualified cases. We now have a situation in which we need to divide infinity by infinity.

Chapter 7: Elementary Probability

When the number of events is uncountable (e.g. infinite), we rely on a countable *proxy*, such as the area in this example. Other common proxies may involve length of a line segment, and measurement of an angle.

Another point to note is that in such problems, the result of *not exceeding* and *less than* may be the same. This may sound counter-intuitive at first. But remember, we are now trying to count the *uncountable* with a countable proxy. In this example, the area is our proxy. A curve which represents the boundary condition (i.e. the distance is equal to half of the radius) has an area of 0. This can provide an intuitive explanation why the phases *not exceeding* and *less than* lead to the same result.

Another common type of problems involving the uncountable is related to real numbers.

Example 7.4.2

Let x and y be two real numbers randomly selected between 0 and 1. What is the probability of $y < x$?

When applying the symmetry argument, you may guess it is $\frac{1}{2}$. This answer is correct. Here, we will demonstrate how to use geometric probability to solve this type of problems.

Solution

Let x and y be coordinates of a point on a coordinate plane. We will then use the area of all qualified points and the area of all possible points as proxies to calculate the probability.

As $0 \leq x \leq 1$, x must lie between the y-axis and line $x = 1$. Likewise, y must lie between the x-axis and line $y = 1$. Therefore the possible points must lie within the unit square as shown below.

Chapter 7: Elementary Probability

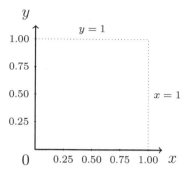

As the question asks for the probability of $y < x$, we draw a line of $y = x$ to represent the boundary condition. All the points that lie below this line satisfy $y < x$. Therefore the qualified area is the shaded triangle as shown below.

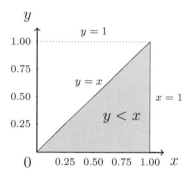

As such, the probability of $y < x$ equals the ratio between the area of the triangle and the area of the square, which is $\frac{1}{2}$.

Done.

Again, you may note that if the question were to ask for the probability of $y \leq x$, the result would be the same as that of $y < x$.

Chapter 7: Elementary Probability

💡 *Tip: Make sure you fully understand this example before continuing.*

Let's recap the key steps in solving such problems.

The first step is to find the region that contains all possible points. The region can be identified on a two-dimensional coordinate grid when there are two random variables, or on a three-dimensional coordinate system when there are three random variables.

The next step is to find the sub-region that contains all qualified points. On a two-dimensional plane, this is usually done by charting one or more lines under the given conditions. On a three-dimensional system, this is usually accomplished by constructing one or more section planes that satisfy the given conditions.

Finally the probability is obtained by calculating the ratio of the two areas (in a two-dimensional case) or the ratio of two volumes (in a three-dimensional case).

We will illustrate with another example:

Example 7.4.3

Let x and y be two real numbers randomly selected between 0 and 1. What is the probability of $x^2 + y^2 < 1$?

Solution

With the same technique and reasoning outlined above, we can conclude the probability equals the quotient of the shared arc area divided by the unit square, which is $\frac{1}{4}\pi$.

Chapter 7: Elementary Probability

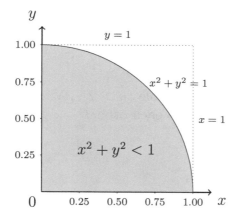

Done.

One remarkable fact shown in this example is that the result contains π. This means that we can use probability to estimate the value of π! If you are interested in exploring further, this will be discussed in *Appendix B* on page 149.

7.5 Practice

Practice 1

n people sit at a round table ($n > 2$). What is the probability that A and B sit next to each other?

Practice 2

Let a and b be two randomly selected points on a line segment of unit length. What is the probability that their distance is not more than $\frac{1}{2}$?

Practice 3

Randomly select 7 letters from "probability" to form a 7-letter word. What is the probability that the 7-letter word is "ability"?

Practice 4

Mary plans to ask Joe to water the flowers during her summer vacation. Joe has a 10% chance of forgetting this chore. If the flowers have an 85% survival rate when watered but only a 20% survival rate when not watered, what is the probability that the flowers will die upon Mary's return?

Practice 5

The germination rates of two different seeds are measured at 90% and 80%, respectively. Find the probability that

(a) both will germinate

(b) at least one will germinate

(c) exactly one will germinate

Practice 6

Joe and Mary plan to meet at the library sometime between 5pm and 6pm. They have agreed that whoever arrives first shall wait for the other person for up to 20 minutes and then leave. What is the probability they will meet?

Practice 7

There are several equally spaced parallel lines on a table. The distance between two adjacent lines is $2a$. On the table, toss a coin with a radius of r ($r < a$). Find the probability that the coin does not touch any line.

Practice 8

Joe breaks a 10-meter long stick into three shorter sticks. Find the probability that these three sticks can form a triangle.

Chapter 7: Elementary Probability

Practice 9

Randomly select 3 real numbers x, y, and z between 0 and 1. What is the probability that $x^2 + y^2 + z^2 > 1$?

Practice 10

In the following diagram, $\overline{AO} = 2$, $\overline{BO} = 5$, and $\angle AOB = 60°$. Point C is selected on $\overline{BO}$ randomly. Find the probability that $\triangle AOC$ is an acute triangle.

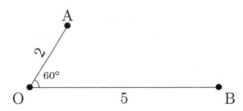

Chapter 7: Elementary Probability

Practice 11

A bug crawls from A along a grid. It never goes backward, it crawls towards all the other possible directions with equal probability. For example:

- At A, it may crawl to either B or D with a 50-50 chance
- At E (coming from D), it may crawl to B, F, or H with a $\frac{1}{3}$ chance each
- At C (coming from B), it will crawl to F for sure

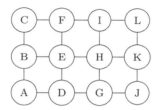

The questions are, from A:

(i) What is the probability of it landing at E in 2 steps?

(ii) What is the probability of it landing at F in 3 steps?

(iii) What is the probability of it landing at G in 4 steps?

Chapter 7: Elementary Probability

Appendices

Appendices

Appendix A

Solutions

Chapter A: Solutions

A.1 *Chapter 1*

This section is intentionally left blank.

So section numbers of solutions and practices can match.

A.2 Chapter 2

Practice 1

Restaurant MAS offers a set menu with 3 choices of appetizers, 5 choices of main dishes, and 2 choices of desserts. How many possible combinations can a customer have for one appetizer, one main dish, and one dessert?

$3 \times 5 \times 2 = 30$.

Practice 2

How many factors does 20 have?

The answer is 6.

This can be done with prime factorization: $20 = 2^2 \times 5$.

Any factor of 20 must be a product of up to two 2s and up to one 5. Therefore we have 3 choices of using 2 (including not using 2) and 2 choices of using 5 (including not using 5). This gives us the answer as $3 \times 2 = 6$.

As a cross check, we can list all the factors manually, as in the following:

factor	# of 2s	# of 5s
1	0	0
2	1	0
4	2	0
5	0	1
10	1	1
20	2	1

Chapter A: Solutions

> 💡 *Tip: In general, if a positive integer N's prime factorization is $N = p_1^{a_1} p_2^{a_2} \cdots p_k^{a_k}$, the count of N's factors is $(a_1 + 1)(a_2 + 1) \cdots (a_k + 1)$.*

> 💡 *Tip: Only perfect squares have an odd number of factors. All other cases must have an even number of factors.*

Practice 3

Find the number of different rectangles that satisfy the following conditions:

- its area is 2015
- the lengths of all its sides are integers

The answer is 4.

$2015 = 5 \times 13 \times 31$. Therefore it has $(1+1) \times (1+1) \times (1+1) = 8$ factors. Because factors appear in pairs, 2015 has 4 pairs of factors. Therefore there are 4 rectangles:

- 1×2015
- 5×403
- 13×155
- 31×65

> ❓ *Quiz: What if the given area is a perfect square with an odd number of factors?*

Practice 4

Eight chairs are arranged in two equal rows. Joe and Mary must sit in the front row. Jack must sit in the back row. How many different seating plans can they have?

Chapter A: Solutions

It does not matter whether these chairs are arranged in one row or two rows. What matters is how many choices each person has.

This problem can be solved in 4 steps:

step 1: Joe selects a seat: 4 choices

step 2: Mary selects a seat: 3 choices

step 3: Jack selects a seat: 4 choices

step 4: Everybody else sits: 5! choices

Therefore the total number of seating plans: $4 \times 3 \times 4 \times 5! = 5760$.

Practice 5

Two Britons, three Americans, and six Chinese form a line:

(a) How many different ways can the 11 individuals line up?

(b) If two people of the same nationality cannot stand next to each other, how many different ways can the 11 individuals line up?

(a) $(2+3+6)! = 11!$

(b) The numbers here are *carefully* selected. Many competition problems are designed in such way. Because we have a total of 5 non-Chinese and 6 Chinese, the only way to satisfy the given condition is to put the non-Chinese into the 5 intervals between the 6 Chinese. Therefore the answer is 5!6!.

Chapter A: Solutions

Practice 6

Use digits 1, 2, 3, 4, and 5 without repeating to create a number.

(a) How many 5-digit numbers can be formed?

(b) How many numbers will have the two even digits appearing between 1 and 5? (e.g.12345)

(a) $5! = 120$.

(b) Let's place 1, 5, 2, and 4 first, as shown in the following. The outside two digits are for 1 and 5 and the inside two digits are for 2 and 4.

$$\underline{}\quad \underline{}\quad \underline{}\quad \underline{}$$
$$\text{1 or 5}\quad \text{2 or 4}\quad \text{2 or 4}\quad \text{1 or 5}$$

There are $(2 \times 1) \times (2 \times 1)$ choices.

After they have been placed, 3 can be positioned with five choices:

Therefore the answer is $(2 \times 1) \times (2 \times 1) \times 5 = 20$.

Practice 7

Joe plans to put a red stone, a blue stone, and a black stone on a 10 × 10 grid. The red stone and the blue stone cannot be in the same column. The blue stone and the black stone cannot be in the same row. How many different ways can Joe arrange these three stones?

The key to solve this problem is to note that the red stone and the black stone have no interdependency. Therefore we can settle them first before placing the more restrictive blue stone.

Chapter A: Solutions

step (1) Choose a position for the red stone: 100 choices

step (2) Choose a position for the black stone: 99 choices

step (3) Choose a position for the blue stone: $9 \times 9 = 81$ choices

Therefore the answer is $100 \times 99 \times 81 = 801900$.

Practice 8

How many different 6-digit numbers can be formed by using digits 1, 2, and 3, if no adjacent digits can be the same?

There are 3 choices for the first digit and 2 choices for all the following digits. Therefore the answer is:

$$3 \times 2 \times 2 \times 2 \times 2 \times 2 = 96$$

Practice 9

How many different ways are there to arrange 3 black balls and 3 white balls in a circle? Two arrangements are considered the same if they are different just by rotating the balls.

The answer is 4.

There are three different cases:

(1) No two black balls are next to each other: 1 possibility

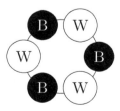

(2) Only two black balls are next to each other: 2 possibilities

Chapter A: Solutions

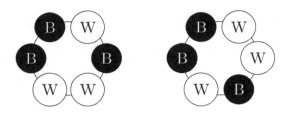

(3) All the three black balls are next to one other: 1 possibility

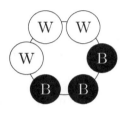

Practice 10

Joe wants to write 1 to n in a $1 \times n$ grid. The number 1 can be written in any grid, while the number 2 must be written next to 1 (can be at either side) so that these two numbers are together. The number 3 must be written next to this two-number block. This process goes on. Every new number written must stay next to the existing number block.

How many different ways can Joe fill this $1 \times n$ grid?

The answer is 2^{n-1}.

This problem can be solved by using backward induction, i.e. start on the last item and then work out the rest backwards one by one till the first number.

We first study possible locations for n. Because the previous $(n-1)$ numbers must stay together, it is obvious that n must be at either the first grid or the last. This means there are 2 choices to place n.

Removing n, by the same reasoning, $(n-1)$ must be at either the first or the last of the $1 \times (n-1)$ grid. This means there are 2 choices to write the number $(n-1)$ as well.

Repeating this process, we find there are also 2 choices to place the number 2. The last remaining grid must contain 1.

Therefore the answer is 2^{n-1}.

To verify the answer, let's use some small numbers:

When $n = 1$, there is obviously only 1 choice. $2^{1-1} = 1$.

When $n = 2$, there are 2 choices. $2^{2-1} = 2$.

When $n = 3$, there are 4 choices: 123, 312, 213, 321. $2^{3-1} = 4$.

Having repeated this checking process, we are confident that we do have the correct answer.

Tip: The ability to check your result is an important and useful skill to develop.

A.3 Chapter 3

Practice 1

Ten people attend a meeting. Each person shakes hands with the other attendees exactly once. How many times do they shake hands in total?

$C_{10}^2 = \frac{10 \times 9}{2} = 45$.

Practice 2

n teams participate in a friendly tournament, in which each team plays against all the other teams exactly once. If there are a total of 36 games played, find n.

n teams will play a total of $C_n^2 = \frac{n(n-1)}{2}$ games. Solving $\frac{n(n-1)}{2} = 36$ gives us $n = 9$.

Practice 3

There are 8 different points on a circle. How many triangles can be formed by using these points as vertices?

$C_8^3 = 56$

Practice 4

How many terms are there if we fully expand the following polynomial?
$$(a_1 + b_1)(a_2 + b_2) \cdots (a_n + b_n)$$

The answer is 2^n.

Each term in the expanded form is a product of n terms, each of which is chosen from one of the n respective brackets. This is equivalent to picking one term from each bracket and multiplying them. Since each bracket offers us 2 choices, n brackets will give us 2^n choices in total.

Practice 5

Eight couples attend a party. Every gentleman will shake hands with all the other guests except his own wife. Ladies do not shake hands with other ladies. Find out the total number of handshakes.

We can use the technique of counting the opposite (complementary counting) to solve this problem.

If 16 people all shake hands with other guests, the total count is $C_{16}^2 = 120$.

If every lady shakes hands with the other ladies, the total count is $C_8^2 = 28$.

If every couple shake hands between themselves, the total count is 8.

Therefore the answer is $120 - 28 - 8 = 84$.

Practice 6

n couples attend a ceremony and sit in a row. If every couple must sit together, how many different seating plans are there?

The answer is $2^n \cdot n!$.

Since every couple must sit together, we can first treat them as a single unit. This leaves us with n units to arrange. The count is $n!$.

Chapter A: Solutions

Then every couple can switch seats between themselves. Therefore n couples have 2^n choices.

By the multiplication principle, the final answer is $2^n \cdot n!$.

Practice 7

In a soccer game, team MAS has 11 players on the field now and 7 substitutes as backup.

Per rules, at most 3 substitutions can be made in any game. A player that is replaced by a substitute cannot go back to the field during that game.

If at the end of this game, team MAS still has 11 players on the field, how many different possibilities are there?

There are four cases:

# of substitutions	possibilities
0	1
1	$C_7^1 C_{11}^1$
2	$C_7^2 C_{11}^2$
3	$C_7^3 C_{11}^3$

Therefore the total possibilities: $1 + C_7^1 C_{11}^1 + C_7^2 C_{11}^2 + C_7^3 C_{11}^3 = 7008$.

A.4 Chapter 4

Practice 1

A group of 30 students participated in after-school sport activities including soccer, basketball, and baseball. Each student participated in at least one activity, but not all three. If 20 signed up for soccer, 12 for basketball, 10 for baseball, 6 for both soccer and basketball, and 2 for both basketball and baseball, how many students signed up for both soccer and baseball?

Let x be the number of students who signed up for both soccer and baseball, we have

$$30 = (20 + 12 + 10) - (6 + 2 + x) + 0$$

Solving this equation gives us $x = 4$.

Practice 2

How many positive, proper fractions are there with a denominator of 2015, in the simplest form?

The answer is 1440.

This problem is equivalent to counting $1 \leq n < 2015$ such that n is relatively prime to 2015.

Let's first count the numbers that are not exceeding 2015 and are NOT relatively prime to 2015. Since $2015 = 5 \times 13 \times 31$, this becomes counting the number of multiples of either 3, 13, or 31, providing that the multiple is not exceeding 2015. There are:

- $\frac{2015}{5} = 403$ multiples of 5
- $\frac{2015}{13} = 155$ multiples of 13

Chapter A: Solutions

- $\frac{2015}{31} = 65$ multiples of 31
- $\frac{2015}{5\times 13} = 31$ common multiples of 5 and 13
- $\frac{2015}{5\times 31} = 13$ common multiples of 5 and 31
- $\frac{2015}{13\times 31} = 5$ common multiples of 13 and 31
- $\frac{2015}{5\times 13\times 31} = 1$ common multiple of 5, 13, and 31

Therefore the count is $(403+155+65)-(31+13+5)+1 = 575$ which means there are $2015-575 = 1440$ numbers that are relatively prime to 2015 (including the case of 1).

This result can also be obtained by using Euler's ϕ function[1]:

$$2015 \times (1 - \frac{1}{5}) \times (1 - \frac{1}{13}) \times (1 - \frac{1}{31}) = 1440$$

Practice 3

Find the number of prime numbers that are not exceeding 100.

The answer is 25.

We first count the number of composite numbers that are not exceeding 100.

Any composite number not exceeding 100 must have at least one prime factor not exceeding $\sqrt{100} = 10$. Therefore, any such composite number must be a multiple of 2, 3, 5, or 7.

Let S_2, S_3, S_5, and S_7 be the sets containing all multiples of 2, 3, 5, and 7, respectively, that are not exceeding 100. We have:

$$|S_2| = \left\lfloor \frac{100}{2} \right\rfloor = 50$$

---

[1]Euler's ϕ function is discussed in the book *Number Theory*.

$$|S_3| = \left\lfloor \frac{100}{3} \right\rfloor = 33$$

$$|S_5| = \left\lfloor \frac{100}{5} \right\rfloor = 20$$

$$|S_7| = \left\lfloor \frac{100}{7} \right\rfloor = 14$$

$$|S_2 \cap S_3| = \left\lfloor \frac{100}{2 \times 3} \right\rfloor = 16$$

$$|S_2 \cap S_5| = \left\lfloor \frac{100}{2 \times 5} \right\rfloor = 10$$

$$|S_2 \cap S_7| = \left\lfloor \frac{100}{2 \times 7} \right\rfloor = 7$$

$$|S_3 \cap S_5| = \left\lfloor \frac{100}{3 \times 5} \right\rfloor = 6$$

$$|S_3 \cap S_7| = \left\lfloor \frac{100}{3 \times 7} \right\rfloor = 4$$

$$|S_5 \cap S_7| = \left\lfloor \frac{100}{5 \times 7} \right\rfloor = 2$$

$$|S_2 \cap S_3 \cap S_5| = \left\lfloor \frac{100}{2 \times 3 \times 5} \right\rfloor = 3$$

$$|S_2 \cap S_3 \cap S_7| = \left\lfloor \frac{100}{2 \times 3 \times 7} \right\rfloor = 2$$

$$|S_2 \cap S_5 \cap S_7| = \left\lfloor \frac{100}{2 \times 5 \times 7} \right\rfloor = 1$$

$$|S_3 \cap S_5 \cap S_7| = \left\lfloor \frac{100}{3 \times 5 \times 7} \right\rfloor = 0$$

$$|S_2 \cap S_3 \cap S_5 \cap S_7| = \left\lfloor \frac{100}{2 \times 3 \times 5 \times 7} \right\rfloor = 0$$

Therefore:
$$|S_2 \cup S_3 \cup S_5 \cup S_7|$$
$$=(50+33+20+14)$$
$$-(16+10+7+6+4+2)$$
$$+(3+2+1+0)$$
$$-0$$
$$=78$$

This means that there are 78 integers that are multiples of 2, 3, 5, or 7. These integers are less than or equal to 100. Among them, 2, 3, 5, and 7 are prime numbers. This leads to 74 composite numbers. We must also exclude 1, which is neither a prime number nor a composite number. Therefore we have $100 - 74 - 1 = 25$ prime numbers.

A.5 Chapter 5

Practice 1

Find the number of positive integers that satisfy both conditions below:

- Is not exceeding 100

- Is a product of 2 prime numbers

The answer is 34.

The smallest prime number is 2. Therefore to make the product not exceeding 100, the other prime number should not exceed 50.

There are altogether 15 prime numbers that do not exceed 50: 2, 3, 5, 7, 11, 13, 17, 19, 23, 29, 31, 37, 41, 43, 47.

Let's count by using the smaller of the two prime factors:

the smaller one	choice of the other	count
2	any of these 15 choices	15
3	3 to 31	10
5	5 to 19	6
7	7 to 13	3
11 or above	no choice	0

Therefore the answer is $15 + 10 + 6 + 3 = 34$.

Practice 2

There are two sets $\{1,3,5,...,23\}$ and $\{2,4,6,...,24\}$. If we select one number from each set and add them together, how many different results can we have in total?

Chapter A: Solutions

We note that the first set contains odd numbers, while the second contains even number. Adding an odd number and an even number results in an odd number. Since the smallest sum is 3 and the largest is 47, the total number of results we can have is $(47-3)/2+1 = 23$, which includes all the odd numbers between 3 and 47, inclusive.

Practice 3

There are 3 red balls and 3 green balls in a jar. If we retrieve one ball at a time, and cannot take out more red balls than green balls at any time, how many different ways are there to retrieve these balls?

solution 1

Clearly, the first one must be a green ball and the last one must be a red ball. We then only have 2 red balls and 2 green balls to arrange in the middle 4 retrievals.

GGGRRR GGRGRR GGRRGR GRGGRR GRGRGR

There are 5 different ways in total.

solution 2

We can also solve this problem with the *lattice* technique. This technique is discussed in *Section 6.3* on *page 56*.

On the following grid, we model by moving a green ball east and a red ball north. Therefore taking 3 balls each is equivalent to moving 3 steps east and 3 steps north on the grid.

For example, the highlighted route is the same as taking GGRRGR, where G stands for a green ball and R stands for a red ball.

As we cannot take more red balls than green balls at any time, we cannot go beyond the dotted line.

Chapter A: Solutions

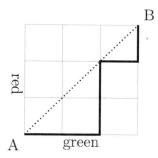

Therefore the original question is equivalent to counting the number of routes from point A to B without going beyond the dotted line.

To count correctly, we will mark each node with the number of routes that can lead to this node. The value of a node equals the sum of the values of the two nodes which lie immediately below and to its left, if they exist.

Therefore the answer is 5. This agrees with the result in Solution 1. The beauty of this solution is that it is unlikely to mis-count.

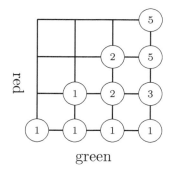

Chapter A: Solutions

Practice 4

Positive integers, such as 12321, are called palindromes because they stay the same if the order of their digits is reversed. Find the number of palindromes that are not exceeding 100,000.

The answer is 1098.

If a palindrome N has k digits, the first $\left\lfloor \frac{k+1}{2} \right\rfloor$ digits will determine this number. Here, $\lfloor x \rfloor$ is the floor operator that returns the biggest integer not exceeding x.

Because the number is not exceeding 100,000, at most it can have 5 digits (e.g. 99,999). We then divide the problem into 5 different cases according to k:

k	$\left\lfloor \frac{k+2}{2} \right\rfloor$	choices
1	1	9
2	1	9
3	2	$9 \times 10 = 90$
4	2	$9 \times 10 = 90$
5	3	$9 \times 10 \times 10 = 900$

Therefore there are $9 + 9 + 90 + 90 + 900 = 1098$ palindromes not exceeding 100,000.

Chapter A: Solutions

Practice 5

Count the number of different ways to distribute 6 concert tickets, seating numbers 1 to 6, among 4 students so that:

- every student gets either 1 or 2 tickets, and

- if 2 tickets are given to the same student, these 2 seats must be next to each other

There are altogether 6 different ways to divide these 6 tickets (see the diagram below) and P_4^4 ways to distribute them among 4 students. Therefore the answer is $6 \times P_4^4 = 144$.

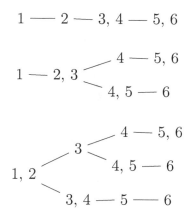

Chapter A: Solutions

A.6 Chapter 6

Practice 1

Seven people form a line. If A must stand next to B, and C must stand next to D, how many possibilities are there?

By applying the *bundling technique*, the answer is $2! \times 2! \times 5! = 480$.

Practice 2

How many triangles are there in this diagram?

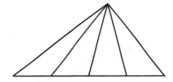

A triangle is defined by 3 vertices. One of them is fixed, therefore this problem is equivalent to *selecting 2 points out of 5 choices, and order is not important*. The answer is $C_5^2 = 10$.

Practice 3

How many rectangles are there in the following diagram?

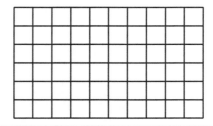

Chapter A: Solutions

Every rectangle in the diagram can be mapped to 4 points, two of which are on the X-axis and the other two are on the Y-axis.

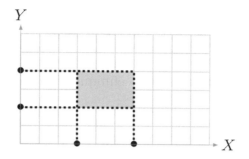

Therefore this problem can be solved in two steps:

step 1: Select 2 out of 11 choices on the X-axis, and order is not important: C_{11}^2

step 2: Select 2 out of 7 choices on the Y-axis, and order is not important: C_7^2

Therefore the answer is: $C_{11}^2 C_7^2 = 1155$.

Practice 4

Team MAS won a total of 10 gold medals in a 6-day tournament. It won at least one gold medal every day. How many different possibilities are there to count the number of gold medals won each day?

According to the *cut the rope* technique, the answer is $C_{10-1}^{6-1} = 126$.

Chapter A: Solutions

Practice 5

Find the number of positive integer solutions to the following equation:
$$x_1 + x_2 + \cdots + x_6 = 20$$

Based on the *cut the rope* technique, the answer as $C_{20-1}^{6-1} = 11628$.

Practice 6

Find the number of non-negative integer solutions to the following equation:
$$x_1 + x_2 + \cdots + x_6 = 20$$

After employing the *knives and balls* technique, we get the answer is $C_{20+6-1}^{6-1} = 53130$.

Practice 7

Joe goes to a supermarket to buy 10 cakes. There are 6 different types of cakes, and each type has a sufficient quantity. How many different combinations of cakes can Joe have?

This is the same as finding the number of non-negative integer solutions to the following equation:

$$x_1 + x_2 + x_3 + x_4 + x_5 + x_6 = 10$$

The answer is $C_{10+6-1}^{6-1} = 3003$.

Chapter A: Solutions

Practice 8

If integers x, y, and z satisfy $0 \leq x \leq 2$, $1 \leq y \leq 3$, and $1 \leq z \leq 2$, how many different combinations of (x, y, z) can satisfy $x + y + z = 4$?

solution 1 - generating function

Expand $(a^0 + a^1 + a^2)(a^1 + a^2 + a^3)(a^1 + a^2)$, we find the coefficient of a^4 is 5. Therefore the answer is 5.

solution 2 - manual count

As the number is small, it is possible to manually count:

x	y	z
0	2	2
0	3	1
1	1	2
1	2	1
2	1	1

Therefore the answer is 5.

Practice 9

A jar contains 5 red balls, 4 green balls, and 3 blue balls. If we take out one ball at a time and record its color, how many different possibilities can we have after all the balls are taken out and their colors are recorded?

This is the *permutation of combination* pattern. Therefore the answer is
$$\frac{(5+4+3)!}{5! \times 4! \times 3!} = 27720$$

Chapter A: Solutions

Practice 10

Seven people with different last names sit at a round table. What is the probability of them sitting in the alphabetical order of their last names either ascendingly or descendingly?

Based on the *circle* pattern, there are $6! = 720$ ways.

Among them, there are exactly 2 ways that satisfy the condition.

Therefore the probability is $\frac{2}{720} = \frac{1}{360}$.

Practice 11

Let integers $1 \leq a_1 < a_2 < a_3 < a_4 < a_5 \leq 25$. The difference between any 2 of these 5 integers is not less than 4. How many different choices of the integers are there?

Because $a_5 \geq a_4 + 4$, we have $a_5 > a_4 + 3$. Similarly we have $a_4 > a_3 + 3$, $a_3 > a_2 + 3$, and $a_2 > a_1 + 3$.

Instead of finding a_5, a_4, a_3, a_2, and a_1, we will try to find a_5, a_4+3, a_3+3, a_2+3, and a_1+3 by determining their possible range.

We have $25 \geq a_5 > a_4 + 3 > a_3 + 6 > a_2 + 9 > a_1 + 12 \geq 13$.

Therefore a_5, $a_4 + 3$, $a_3 + 3$, $a_2 + 3$, and $a_1 + 3$ are 5 different integers between 13 and 25, inclusive. This is equivalent to *selecting 5 different integers from 13 to 25, and order is not important*.

A selection of these 5 integers will uniquely determine a_5, a_4, a_3, a_2, and a_1. Therefore the answer is $C_{13}^5 = 1287$.

Chapter A: Solutions

Practice 12

To arrange 3 white balls and 8 black balls in a circle so that every two white balls must be separated by at least two black balls. How many different arrangements do we have? (Two arrangements will be considered the same if they are different just by rotating the balls.)

The answer is 2.

Based on the technique presented in *Example 6.4.2* on *page 59*, we can bundle a white ball with two neighboring black balls as a super ball. Now the problem is simplified to arranging 3 super balls and 2 remaining black balls in a circle. There are only 2 possibilities to arrange them. (Do not count mirroring arrangements!)

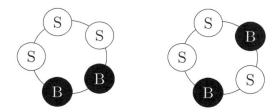

Practice 13

Find the number of different ways to assign 5 different tasks to 4 workers so that every worker gets at least one task.

According to the conclusion from *Section 6.8* on *page 67*, the answer is:
$$4^5 - 3^5 C_4^1 + 2^5 C_4^2 - 1^5 C_4^3 = 240$$

Chapter A: Solutions

Practice 14

Four people attending a gathering bring a gift each for exchange. Everybody shall get one gift that is not his own. How many different ways are there to exchange these gifts?

It is a derangement problem. Therefore the answer is

$$D_4 = 4! \times (\frac{1}{2!} - \frac{1}{3!} + \frac{1}{4!}) = 9$$

A.7 Chapter 7

Practice 1

n people sit at a round table $(n > 2)$. What is the probability that A and B sit next to each other?

Solution 1

For n people to sit at a round table, there are $(n-1)!$ possibilities.

If A and B must sit together, there are $2 \times (n-2)!$ possibilities according to the bundling technique

Therefore the answer is $\frac{2 \times (n-2)!}{(n-1)!} = \frac{2}{n-1}$.

Solution 2

Let A be seated first. Then there are only 2 seats next to A for B to choose from the remaining $(n-1)$ seats. Therefore the probability is $\frac{2}{n-1}$.

Practice 2

Let a and b be two randomly selected points on a line segment of unit length. What is the probability that their distance is not more than $\frac{1}{2}$?

This problem is equivalent to selecting two real numbers, x and y, between 0 and 1 so that $|y - x| \leq \frac{1}{2}$.

The region that contains all possible points is still the unit square.

$|y - x| = \frac{1}{2}$ is equivalent to $y - x = \pm \frac{1}{2}$, which can be re-arranged

Chapter A: Solutions

as

$$\begin{cases} y = x + \frac{1}{2} \\ y = x - \frac{1}{2} \end{cases}$$

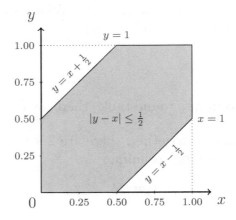

Therefore the qualified region is the shaded area as shown. This gives us an answer of $\frac{3}{4}$.

> *Tip: In order to determine which one is the qualified region, simply pick a point to test. For example, in this case, we select the center point (0.5,0.5) which clearly satisfies the condition. However the corner point (1,0) does not meet the condition. As a result, we are sure the shaded region is the qualified region.*

Practice 3

Randomly select 7 letters from "probability" to form a 7-letter word. What is the probability that the 7-letter word is "ability"?

We can treat the two "i"s and two "b"s as distinguishable. Therefore there are a total of 4 cases in which the 7 letters can form the word "ability".

Chapter A: Solutions

There are P_{11}^7 possibilities to select 7 letters from 11 choices in order to form a line.

Therefore the answer is $\frac{4}{P_{11}^7}$.

Practice 4

Mary plans to ask Joe to water the flowers during her summer vacation. Joe has a 10% chance of forgetting this chore. If the flowers have an 85% survival rate when watered but only a 20% survival rate when not watered, what is the probability that the flowers will die upon Mary's return?

Let's construct a probability tree diagram:

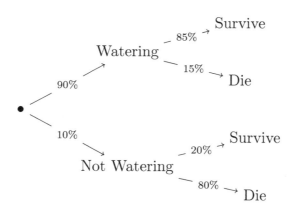

Therefore the answer is $90\% \times 15\% + 10\% \times 80\% = 21.5\%$.

Chapter A: Solutions

Practice 5

The germination rates of two different seeds are measured at 90% and 80%, respectively. Find the probability that

(a) both will germinate

(b) at least one will germinate

(c) exactly one will germinate

(a) $90\% \times 80\% = 72\%$

(b) $90\% + 80\% - 90\% \times 80\% = 98\%$, or $1 - (1 - 90\%) \times (1 - 80\%) = 98\%$

(c) $90\% \times (1 - 80\%) + 80\% \times (1 - 90\%) = 26\%$

Practice 6

Joe and Mary plan to meet at the library sometime between 5pm and 6pm. They have agreed that whoever arrives first shall wait for the other person for up to 20 minutes and then leave. What is the probability they will meet?

This problem is equivalent to:

Let x and y be two random variables between 0 and 1. Find the probability that $|x - y| \leq \frac{1}{3}$.

The fraction $\frac{1}{3}$ comes from the fact that 20 minutes is $\frac{1}{3}$ of an hour (i.e. between 5pm and 6pm).

By using the techniques presented earlier, we can conclude the answer as $\frac{5}{9}$.

Chapter A: Solutions

Practice 7

There are several equally spaced parallel lines on a table. The distance between two adjacent lines is $2a$. On the table, toss a coin with a radius of r ($r < a$). Find the probability that the coin does not touch any line.

The answer is $\frac{a-r}{a}$.

It is clear from the diagram below that the coin will not touch the lines if and only if its center falls within the two dashed lines.

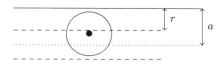

Practice 8

Joe breaks a 10-meter long stick into three shorter sticks. Find the probability that these three sticks can form a triangle.

The answer is $\frac{1}{4}$.

Let x and y be the lengths of the first two sticks, respectively. Then the length of the third stick is $10 - x - y$. Therefore we have the qualified region defined by:

$$\begin{cases} 0 < x < 10 \\ 0 < y < 10 \\ 0 < 10 - x - y \end{cases}$$

This is shown in the following diagram:

Chapter A: Solutions

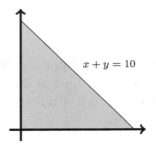

In order to form a triangle, the following relationships must hold:

$$\begin{cases} x + y > 10 - x - y \\ x + (10 - x - y) > y \\ y + (10 - x - y) > x \end{cases}$$

These define three boundary conditions:

$$\begin{cases} x + y = 5 \\ y = 5 \\ x = 5 \end{cases}$$

Therefore, the qualified region shaded below shows the final answer as $\frac{1}{4}$.

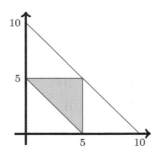

Chapter A: Solutions

Practice 9

Randomly select 3 real numbers x, y, and z between 0 and 1. What is the probability that $x^2 + y^2 + z^2 > 1$?

This problem is similar to *Example 7.4.3* on *page 102*, although it is in a 3-D space.

The qualified region is the place that lies outside the sphere, but inside the unit cubic. Therefore the result is:

$$1 - \frac{1}{8}(\frac{4}{3}\pi) = 1 - \frac{1}{6}\pi$$

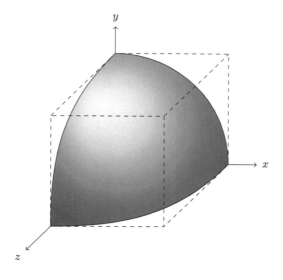

Chapter A: Solutions

Practice 10

In the following diagram, $\overline{AO} = 2$, $\overline{BO} = 5$, and $\angle AOB = 60°$. Point C is selected on $\overline{BO}$ randomly. Find the probability that $\triangle AOC$ is an acute triangle.

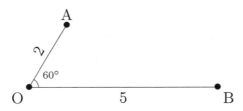

Let's start with the boundary situation: when $\triangle AOC$ is a right triangle.

Draw line $\overline{AC_1} \perp \overline{OB}$ and line $\overline{AC_2} \perp \overline{AO}$ where both C_1 and C_2 are points on OB. Based on the $30° - 60° - 90°$ right triangle property, $\overline{OC_1} = 1$ and $\overline{OC_2} = 4$.

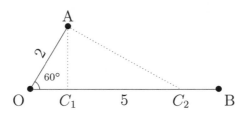

Clearly if point C falls to the left of C_1, $\angle ACO$ will be an obtuse angle. If C falls to the right of C_2, $\angle OAC$ will be an obtuse angle. If and only if C falls between C_1 and C_2 will $\triangle AOC$ be an acute triangle.

Therefore the answer is $\frac{\overline{C_1 C_2}}{\overline{OB}} = \frac{3}{5}$.

Chapter A: Solutions

💡 *Tip: This example also shows the importance and effectiveness of working with boundary conditions.*

Practice 11

A bug crawls from A along a grid. It never goes backward, it crawls towards all the other possible directions with equal probability. For example:

- At A, it may crawl to either B or D with a 50-50 chance

- At E (coming from D), it may crawl to B, F, or H with a $\frac{1}{3}$ chance each

- At C (coming from B), it will crawl to F for sure

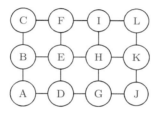

The questions are, from A:

(i) What is the probability of it landing at E in 2 steps?

(ii) What is the probability of it landing at F in 3 steps?

(iii) What is the probability of it landing at G in 4 steps?

(i) The bug can travel from A to E in 2 steps along two paths:

Chapter A: Solutions

$$A \begin{matrix} \nearrow^{\frac{1}{2}} B - \frac{1}{2} \to E \\ \searrow_{\frac{1}{2}} D - \frac{1}{2} \to E \end{matrix}$$

Therefore, the probability is

$$\frac{1}{2} \times \frac{1}{2} + \frac{1}{2} \times \frac{1}{2} = \frac{1}{2}$$

(ii) The bug can move from A to F in 3 steps along three paths.

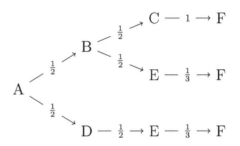

Therefore, the probability is

$$\frac{1}{2} \times \frac{1}{2} \times 1 + \frac{1}{2} \times \frac{1}{2} \times \frac{1}{3} + \frac{1}{2} \times \frac{1}{2} \times \frac{1}{3} = \frac{5}{12}$$

(iii) Three paths can lead the bug from A to G in 4 steps.

$$A \begin{matrix} \nearrow^{\frac{1}{2}} B \begin{matrix} \nearrow^{\frac{1}{2}} E - \frac{1}{3} \to H - \frac{1}{3} \to G \\ \searrow_{\frac{1}{2}} E - \frac{1}{3} \to D - \frac{1}{2} \to G \end{matrix} \\ \searrow_{\frac{1}{2}} D - \frac{1}{2} \to E - \frac{1}{3} \to H - \frac{1}{3} \to G \end{matrix}$$

Therefore, the probability is

$$\frac{1}{2} \times \frac{1}{2} \times \frac{1}{3} \times \frac{1}{3} + \frac{1}{2} \times \frac{1}{2} \times \frac{1}{3} \times \frac{1}{2} + \frac{1}{2} \times \frac{1}{2} \times \frac{1}{3} \times \frac{1}{3} = \frac{7}{72}$$

Done.

Appendix B

Estimate π

We have mentioned in *Example 7.4.3* on *page 102* that it is possible to estimate π by applying probability. In fact, some people have tried something similar. One of the most famous examples is *Buffon's Needle*. You can google it on the internet. Its formula may be beyond your comprehension, but the fundamental idea is the same.

It is remarkable that π can be estimated by dropping a needle, isn't it?

We can also utilize the conclusion of *Example 7.4.3* to estimate π. This involves randomly generating two real numbers between 0 and 1 and checking whether the sum of their squares is less than 1. This can be simulated with a spreadsheet.

If you want to see it in action, you can download the spreadsheet, EstimatePi.xls, from:

http://www.mathallstar.com/Files/Supplementary

You will need Excel (or OpenOffice) to open this spreadsheet. As it contains macros in order to simulate the experiment, Excel

Chapter B: Estimate π

may warn you as a security precaution. Simply click the "enable macro" button.

Results will vary between different experiments due to the random nature in selecting two real numbers. In general, the more tries you model, the closer the estimate is to the value of π.

Below is the result of a sample experiment:

Tries	Probability	Estimation
1	0.000000	0.000000
10	0.500000	2.000000
100	0.740000	2.960000
1,000	0.794000	3.176000
10,000	0.786500	3.146000

Estimate

Made in United States
Troutdale, OR
10/27/2023